Hope Amidst Conflict

SERIES IN POLITICAL PSYCHOLOGY

Image Bite Politics: News and the Visual Framing of Elections
Maria Elizabeth Grabe and Erik Page Bucy

Social and Psychological Bases of Ideology and System Justification
John T. Jost, Aaron C. Kay, and Hulda Thorisdottir

The Political Psychology of Democratic Citizenship
Eugene Borgida, Christopher M. Federico, and John L. Sullivan

On Behalf of Others: The Psychology of Care in a Global World
Sarah Scuzzarello, Catarina Kinnvall, and Kristen R. Monroe

The Obamas and a (Post) Racial America?
Gregory S. Parks and Matthew W. Hughey

Ideology, Psychology, and Law
Jon Hanson and John Jost

The Impacts of Lasting Occupation: Lessons from Israeli Society
Daniel Bar-Tal and Izhak Schnell

Competing Motives in the Partisan Mind
Eric W. Groenendyk

Personalizing Politics and Realizing Democracy
Gian Vittorio Caprara and Michele Vecchione

Representing Red and Blue: How the Culture Wars Change the Way Citizens Speak and Politicians Listen
David C. Barker and Christopher Jan Carman

The Ambivalent Partisan: How Critical Loyalty Promotes Democracy
Howard G. Lavine, Christopher D. Johnston, and Marco R. Steenbergen

Disenchantment with Democracy: A Psychological Perspective
Janusz Reykowski

Divided: Open-Mindedness and Dogmatism in a Polarized World
Edited by Victor Ottati and Chadly Stern

Hope Amidst Conflict: Philosophical and Psychological Explorations
Oded Adomi Leshem

Hope Amidst Conflict

Philosophical and Psychological Explorations

Oded Adomi Leshem

OXFORD
UNIVERSITY PRESS

Oxford University Press is a department of the University of Oxford. It furthers the University's objective of excellence in research, scholarship, and education by publishing worldwide. Oxford is a registered trade mark of Oxford University Press in the UK and certain other countries.

Published in the United States of America by Oxford University Press
198 Madison Avenue, New York, NY 10016, United States of America.

Library of Congress Cataloging-in-Publication Data
Names: Leshem, Oded Adomi, author.
Title: Hope amidst conflict : philosophical and psychological
exploration / Oded Adomi Leshem.
Description: New York, NY : Oxford University Press, [2024] |
Includes bibliographical references and index.
Identifiers: LCCN 2023017310 (print) | LCCN 2023017311 (ebook) |
ISBN 9780197685303 (hardback) | ISBN 9780197685327 (epub) |
ISBN 9780197685334 (online)
Subjects: LCSH: Conflict management. | Hope. | Social conflict. |
War. | Peace-building.
Classification: LCC HM1126 .L4425 2024 (print) | LCC HM1126 (ebook) |
DDC 303.6/9—dc23/eng/20230515
LC record available at https://lccn.loc.gov/2023017310
LC ebook record available at https://lccn.loc.gov/2023017311

DOI: 10.1093/oso/9780197685303.001.0001

Printed by Integrated Books International, United States of America

This book is dedicated to the children living in the dire reality of conflict and to those working to hasten peace for their sake.

Contents

Foreword

The concept of hope is wide and broad and depends on context, culture, and language. Philosophers, neuroscientists, psychologists, linguists, political scientists, and anthropologists have added their share to its variations. Oded Adomi Leshem's book outlines these meanings and wisely distills from these meanings a novel conceptualization of hope that is both exhaustive and concise. This conceptualization, which is termed the "bidimensional model of hope," is then used throughout the book in numerous studies carried out in the context of conflict and peace. From qualitative interviews with peace activists on the meaning-making of hope to quantitative linguistic analyses of dozens of speeches made by leaders, the book's creative research approach demonstrates the exceptional utility of the *bidimensional model of hope*. The model's versatility and relevance are also exemplified in large-scale studies conducted in Israel, the West Bank, and the Gaza Strip on the role of hope amidst conflict. These are all major contributions to understanding hope in a unitary way.

Leshem's discussion of hope begins with the fantastic tale about hope's journey as told in Greek mythology. According to the myth of Pandora's Box, hope was instilled in humans to let them yearn for something better than their mundane lives, which were full of troubles and worries. Indeed, hope emerges strongest in times of grief and hardship. Thus, the beginning of hope in the case of conflicts stems from the sorrow arising from human causalities, economic loss, the trauma of violence, and the misery of living in constant stress.

Hope is the impetus to start a new phase of the conflict, which can lead to peacebuilding. The road to peacebuilding begins when at least a few members of society come to strongly desire peace and act to realize this idea. And here is the main point: peacebuilding is not wishful thinking; it is the concrete result of a deliberative process. This is a crucial distinction because most people wish for peace. But an empty wish leads people to engage in utopian imagery with abstract ideas that only decrease the likelihood of achieving them. Dreaming is only half of hoping—it is like half a promise that will only happen if humans work hard to realize the dream.

Hope thus consists of concrete positive goals and, in cases of conflict, longing for relief from the terrible situation of intractability. Accompanied by an emotional reaction, this hope requires positively valued, realistic, and concrete goals and directed thinking with pragmatic ways to achieve those goals. Hope is based on higher cognitive processing of mental representations and, more specifically, on setting goals, planning, use of imagery, creativity, cognitive flexibility, mental exploration of novel situations, and risk-taking. Hope can be seen as a state of mind that requires the development of new "scripts": plans for future action.

Leshem's examination of hope goes beyond the intuitive as he takes the reader into underexplored territories, the first of which is a bold look into the dangers and drawbacks of hope. Hope can distort people's ability to see unpleasant reality for what it is, lead to anguish and deep frustration when it is dashed, or, in some cases, lull people into passiveness and compliance. Thus, alongside its merits, hope has pitfalls we should identify and learn to avoid. The book skillfully illustrates the perils and advantages of hope and outlines a way that hope's disadvantages can be minimized and its benefits exploited. As the book reveals, the argument for this approach to hope is not only backed by Leshem's data but can also be found in the work of philosophers, researchers, and politicians.

Sustaining hope is vital during conflict. It is hope that liberates people from their fixed beliefs about the irreconcilability of the conflict, and it is hope that sends them to find creative ways of resolving it. Hope enables people to imagine a future that is different from the past and the present and motivates society members to change their situation by performing acts that were once unthinkable. Without hope, it is impossible to embark on the road to peace. But how can we instill hope among people mired in decades of violence and grief? The book answers this question by detailing several hope-inducing interventions Leshem and others designed and tested. Scholars and practitioners will find these interventions enlightening. As the studies show, hope can be induced, even among the most skeptical, and, in some cases, revive the idea of peace.

Once such an idea emerges and is propagated by society members, a process of mobilizing the rest of society to resolve the conflict begins. It is a long process of societal change that involves building a new repertoire to reach an agreement with the adversary and construct a new ethos that serves as the foundation of a culture of peace. It is a process that involves all members of society, from the grassroots to leaders, and its success depends on a shift in the behavioral repertoire. In this process, society members must alter their basic premises, assumptions, or aspirations—in fact, they must change the worldview that has dominated their lives for many years. Ideologies, cultures, and identity-related beliefs are powerful forces inhibiting this necessary transformation. Moreover, even reaching a peaceful settlement does not necessarily indicate the end of the peacebuilding process.

Through this entire process, it is necessary to embrace hope. You can change the plans, you can change the tactics, or you can change elements of the promises, but you cannot stop hoping. Indeed, the journey of hope is bound to face enormous difficulties. Hope must overcome great barriers that threaten to prevent it from reaching its true potential. Yet only continuous adherence to hope may achieve the desired peace. This book is one of the few attempts—and maybe the only attempt—to reveal the importance of hope in intractable conflict. It is a book for which we have waited a long time.

—Daniel Bar-Tal

Preface

I first met Ibtisam in 2007, when I began volunteering as a driver in a grassroots network of Israeli activists driving sick Palestinian children to hospitals in Israel. Ibtisam, a forty-year-old Palestinian woman from Gaza, was accompanying her ten-year-old nephew Ahmad who was scheduled for an urgent bone marrow transplant. For lack of adequate medical facilities in Gaza, Ahmad's sister had died from the same blood disease two years earlier. Now, Ibtisam was doing everything she could to save her nephew.[1]

Following the successful medical operation, Ahmad was given a demanding schedule of checkups slated three months apart. Missing appointments could cost him his life. From 2008 to 2013, I made about twenty trips from Erez Crossing on the border of the Gaza Strip to the Israeli hospital where Ahmad was being treated. The long hours on the road gave Ibtisam and me ample time to talk about life amidst the conflict. Ibtisam would describe in detail the horrific situation in the Strip, and I would update her about developments in Israeli and international politics. We soon discovered that we shared the same concerns for the future of the people living between the River and the Sea. We quickly became friends.

Our friendship endured the violent clashes that came and went. I particularly remember the weeks in December 2008 when the Israeli Defense Forces (IDF) struck the Gaza Strip with massive force, and Hamas fired hundreds of missiles from the Strip into Israel. When missiles hit Tel-Aviv, Ibtisam called to check if we were all right. Of course, she needed to endure much crueler conditions. Unlike my family and me, she and her loved ones could not find safety in bomb shelters. She had witnessed bombs exploding above her head and amputated bodies lying in the streets outside her door. Yet, when I drove Ibtisam and Ahmad to the hospital some months later, she was hopeful, "Peace is bound to come," she said, "and we must hasten its arrival." I certainly agreed. However, I could not understand how she could have hope for peace while living in such devastating circumstances.

Ibtisam's uncompromising hope amid unbelievable pain sparked my curiosity about hope and its role during conflict. Her approach reminded me of Victor Frankl's notion about the necessity of hope in our lives. Frankl noted

[1] Names were altered to ensure anonymity.

that hoping that the nightmare in the Nazi camps would end, maybe today, maybe tomorrow, was necessary for survival. Those who ceased to hope perished the next day. Hope's perplexing relationship with reality is one of the features I have researched since I began my inquiry into hope. I have also embarked on a journey to find an accurate and comprehensive way to conceptualize hope and use this conceptualization to measure hope for peace in conflict zones worldwide. I also became extremely interested in philosophical explorations of hope and in the study of hope among political elites. Since I started my inquiry into hope, my interest in this elusive concept has not stopped growing. In fact, it has been multiplying as my investigations got deeper and broader.

At first glance, hope might seem to be a spiritual matter, a construct too fuzzy for empirical scrutiny, "a thing with feathers," as Emily Dickenson famously wrote. However, as I soon discovered, philosophers have had a vast interest in hope's features and meanings. Their work has been followed by another strand of inquiry conducted in the second half of the past century by psychologists who were eager to reveal the psychological mechanisms of hope. During the same time, political theorists were trying to define and describe the politics of hope and skepticism. Finally, in the past decade, experimental research in political psychology, including my own, has rediscovered hope as a powerful construct that may facilitate conflict resolution and social change. Surprisingly, these impressive bodies of knowledge have never been in dialogue, their insights never aggregated to create a fuller picture of hope's role in politics and conflict.

Hope Amidst Conflict aims to do just that. In the past years, I have been connecting the dots between studies from these rich and diverse scholarly projects to generate more nuanced and generalizable conclusions about hope and its function during prolonged conflict. Long-lasting violent conflicts push hope for peace to its limits and thus provide a unique setting not only for theoretical examination but also for empirical scrutiny. The more I read, the more it became apparent that the wisdom from the different disciplines must be merged. The more I scrutinized, the clearer became the necessity to rectify the confusion in the literature on hope during conflict (a confusion that also affected my own work). The more I researched, the more it became evident that there was a real need for a comprehensive project on hope amidst conflict.

I HOPE you enjoy the read.

Acknowledgments

It is a pleasure to acknowledge the many people who assisted me in the long but highly rewarding process of writing this monograph.

I first want to thank my colleagues from the Carter School for Peace and Conflict Resolution: Kevin Avruch, Jim Witte, Richard Rubenstein, Douglas Irvin-Erickson, and especially Thomas Flores, who encouraged my curiosity about hope and supported my endeavor to study hope amidst conflict throughout my PhD studies.

I am also in great debt to all those who helped raise the resources for the Israel-Palestinian Hope Map Project. Running such an elaborate surveying apparatus in a conflict zone necessitated the cooperation of many individuals working in Israel, the Occupied Palestinian Territories, and Washington, DC. It also required a relatively big budget. I am therefore grateful for Ms. Mehra Rimer and the dozens of contributors who participated in the successful crowdfunding campaign that enabled the implementation of the Hope Map Project. A special appreciation goes to my friend and colleague Obada Shtaya and the Palestinian researchers who collected the data in the West Bank and Gaza Strip (and wished to remain anonymous). Their high professional standards and dedication brought to the frontline of scholarship the voices and hopes of Palestinians living in the Occupied Palestinian Territories.

Much appreciation goes to my many colleagues, professors and students alike, who devoted their time to think with me, sometimes for hours on end, about hope, conflict, and Israel-Palestine. Acknowledgments go to Boaz Hameiri, Deborah Shulman, Eric Shuman, Orly Idan, Siwar Hasan-Aslih, Ruthie Pliskin, Smadar Cohen-Chen, Oliver Fink, Steven van den Heuvel, Uriel Abulof, Sami Adwan, Neta Oren, Erik Olsman, Nimrod Rosler, Beatrice Hasler, Maor Shani, Aviad Rubin, and others who inspired my thoughts and writing.

A special thank you is due to the seven extremely kind colleagues who mercilessly peer-reviewed the book's chapters: Eran Halperin, Amit Goldenberg, Deborah Shulman, Noa Schori-Eyal, Julia Chaitin, Lior Lehrs, and Anna Stefaniak. Their wise comments and suggestions were essential in helping me justify and clarify my thoughts. I also wish to express my gratitude to Vered Vinitzky-Seroussi from the Harry S. Truman Institute, who hosted me as a postdoc while writing the manuscript, and to Danny Bar-Tal, who, apart from writing the book's foreword, encouraged me to embark on the demanding and gratifying journey of writing *Hope Amidst Conflict.*

I feel delighted to acknowledge the tremendous contribution of my diligent students Shanny Talmor, Ilana Ushomirsky, and Emma Paul, whose hard work made this book possible. Shanny's eye-opening interviews with Palestinian and Israeli peace activists form the basis of Chapter 5, while Ilana's thought-provoking analysis of speeches made by Israeli and Palestinian leaders lies at the heart of Chapter 6. Emma's significant contribution to the stylistic and technical aspects of the book is apparent throughout the manuscript. Most notably, the contribution of these talented young scholars is not merely technical. Shanny, Ilana, and Emma led central research endeavors uncovering the role of hope during conflict. It has been a great privilege to work with them and watch them mature and develop.

It would have been impossible to write this book without the invaluable support, involvement, and guidance of Eran Halperin, the head of the PICR Lab at the Hebrew University. Always available to offer his excellent advice, Eran's insightful feedback (from micro-level editing suggestions to macro-level strategizing) was always on the mark. Eran was also generous in providing financial resources to aid the book's timely progress. Yet, as my postdoctoral adviser, Eran's contribution to this book goes well beyond advice and resources. His mentorship has made me the scholar I am, and for this, I am genuinely grateful.

Last, I am immensely thankful for the endless love and encouragement from my parents Benny and Haya, my lovely wife Yasmin, and my charming kids, Aviv, Maayan, and Nadav.

1
Introduction—Linking Conflict with Hope

> Hope requires the conviction about the yet-unproven.
>
> —**Eric Fromm**

What is the role of hope in the lives of individuals and collectives mired in conflict and political unrest? How do political elites utilize hope and skepticism to mobilize citizens, and how does hope manifest itself at the grassroots level? Under what conditions can hope proliferate in the seemingly hopeless situations of intractable conflicts, and why do despair and fear often prevail? Is being hopeful necessarily a good thing, or are there situations when hope should be avoided, or at least minimized? These questions are not only thought-provoking but also highly relevant to the reality of millions of people around the world struggling for justice, equality, and peace.

Of course, the question of hope is not confined to the context of conflict. It is located at the center of one of the biggest dilemmas of human survival. What should we accept, and what should we hope to transform? When shall we adapt to reality, and when should we challenge it? The evolution of humankind owes its success to both survival strategies (Hecht, 2013). On the one hand, humans' unique ambition to challenge difficult circumstances and hope for grandiose goals has led humankind to thrive and succeed. On the other hand, humans' adaptivity and capacity to endure rather than challenge difficulties have been pivotal to humankind's survival and prosperity. How are we to choose between these two opposite praxes? In his seminal book *The Critique of Pure Reason*, Immanuel Kant included this dilemma as one of the three basic questions we must all ask: "What may we hope for?" (and, by extrapolation, "What we may not?").

In the context of conflict and political turmoil, the dilemma of hope can be exemplified in the debate between Milan Kundera and Vaclav Havel, two prominent Czechoslovakian writers who found themselves involved in

Hope Amidst Conflict. Oded Adomi Leshem, Oxford University Press. © Oxford University Press 2024.
DOI: 10.1093/oso/9780197685303.003.0001

politics in the late 1960s, when the Soviet Union invaded Czechoslovakia. The Soviets, together with other allies from the Warsaw Pact, sought to crush the Czechoslovakian reform movement that was gaining traction in and outside Prague. Kundera and Havel, the former a successful novelist, the latter a renowned playwright, spoke extensively on politics and the predicament of Czechoslovakians. Yet the two expressed opposite approaches regarding the recommended path their people should choose.

Kundera argued for accommodation, talking about the unpleasant political circumstance as the "fate" or "lot" of the Czechoslovakian people (Sabatos, 2008). For their own survival, claimed Kundera, Czechoslovakians should focus on adapting to the situation rather than challenging it. Indeed, the Czechoslovakians were feeble compared to the mighty USSR and its allies. Perhaps realizing that the odds were so skewed in favor of the Soviets, Kundra leaned toward adjustments and reform, not confrontation or revolution. In Kundra's view, hoping otherwise was a dangerous delusion that would harm the Czechoslovakian people.

Vaclav Havel, his seven-year-younger colleague, accused Kundera of succumbing to a false fate of submission instead of hoping for a future free from foreign influence. Havel insisted that hope should lead the Czechoslovakians in these turbulent times instead of Kundera's pessimism posing as pragmatism. After the invasion, Havel talked about hope for freedom in secret gatherings and wrote about hope in underground magazines. Yet, interestingly, Havel insisted that this hope had nothing to do with the prediction that Czechoslovakians could drive out Soviet influence (Havel, 1990). Speaking about the necessity to free his country from rigid communist dogmas, Havel dismissed the need to base hope on estimations of success. "Hope is not the conviction that something will turn out well, but the certainty that something makes sense, regardless of how it turns out" (p. 181).

Havel thus locates hope in humans' desires and willpower rather than in the estimations that this willpower would achieve its goal. "Hope is a dimension of the soul," he claimed, "not essentially dependent on some particular observation of the world or estimate of the situation" (Havel, 1990, p. 181). For Havel, hope was about aspirations and desires, not predictions and assessments. Havel was imprisoned for his political opinions and "transgressive" activities. Eventually, after twenty years of hope-inspired action, the Havel-led popular struggle paid off. In 1989, Czechoslovakia restored its democracy, and Havel became its first democratically elected president.

When it comes to the hope to withstand Moscow's aggression, Ukrainian president Volodymyr Zelenskyy is perhaps Havel's 21st-century successor. Since the first day of Russia's invasion of Ukraine in February 2022, Zelenskyy expressed high hopes that the Ukraine people would prevail against the much stronger Russian forces. However, the hope expressed by Zelenskyy was quite different from Havel's hope. Havel's hope was not about the estimation of success, whereas Zelenskyy's hope was. "Today, we still believe in the new victory of Ukraine, and we are all convinced that we will not be destroyed by any horde or wickedness," promised Zelenskyy to his people and the international community in his speech delivered two months after the invasion. In other words, for Zelenskyy, hope is not only about willpower but also about the confidence this willpower will deliver.

As of spring 2023 the time of the writing of this book, it is impossible to know if Zelenskyy's hope will be fulfilled. Western culture and modern viewpoints tend to believe that hope always prevails and that people's stubborn desires, positive expectations, and uncompromising commitment to social and political change ultimately pay off. Yet history does not always lean in the direction of the hopeful. Looking at the devastation and brutality unleashed against Syrian men and women hoping to gain basic political freedoms paints hope in much darker colors. More than ten years of cruel civil war left, as of 2022, more than half a million dead, 6.9 million displaced, and all the others utterly hopeless.[1] Political processes are sometimes steered by people's hope and sometimes clash against it.

This book examines hope in extreme political conditions. More specifically, it focuses on the role of hope in intractable violent conflicts—those enduring international disputes deemed by many as unresolvable. Three foundational research questions guide the book's examination of hope amidst intractable conflict. The first question is: "What is hope?" At first glance, the question might appear superfluous. Everybody knows what hope is. Nonetheless, as thoroughly elaborated throughout the book (in particular, Chapter 3), laypeople and experts define and interpret hope in several, sometimes contradicting ways. If we seek to understand hope amidst conflict, we first must agree on its definition. Luckily, philosophers, psychologists, and other hope scholars have debated extensively the question of hope. To understand what hope "is," scholars have looked inward—at hope's components, anatomy, and properties. To grasp what hope "means," they have looked outward—at hope's functions and roles in people's lives. This book combines these two directions to create a holistic and standardized way to understand hope. This understanding informs the way I conceptualize and operationalize hope, which is then used throughout the book.

[1] https://gho.unocha.org/syrian-arab-republic

The second foundational question concerns the determinants of hope in times of political crisis. Here I ask: "What predicts hope during long-standing intractable conflict? Are the young more hopeful than the old? Are the religious more hopeful than the secular? Is political ideology associated with hope, and if so, how?" I also place hope in the context of asymmetrical conflict and ask: "Who hopes for peace more, members from the disadvantaged group who commonly endure harsh, sometimes brutal, experiences of conflict, or those from the advantaged group whose experiences of conflict are usually more moderate?" I also explore behavioral factors that might predict hope (or hopelessness) in intractable conflicts, namely, whether actively working for peace generates hope or hinders it. Identifying hope's determinants is critical for both theoretical and applied purposes. Theories of intractability would benefit from adding a novel lens regarding hope's political and sociopsychological origins. In terms of applied work, looking at the antecedences of hope could also help peacebuilders working in intractable disputes to effectively generate and sustain hope amidst conflict and turmoil.

The third and last foundational question this book poses, is whether hope matters. More specifically, I ask whether hope has consequences for conflict transformation. This question must be seriously examined to build effective strategies for resolving conflicts. If hope generates attitudes and behaviors more conducive to conflict resolution, increasing hope can positively impact peoples' support for policies and actions that advance conflict transformation and peace. If, on the other hand, hope has no impact on conflict-related attitudes and behaviors, then this path seems less promising for promoting the resolution of conflicts. We should also keep in mind that hope may have negative consequences, for example, on people's ability to endure prolonged and violent disputes. In these difficult circumstances, hopes may be quickly shattered, leaving citizens in deeper anguish and despair. Other negative ramifications of hope should also be considered. The question of hope's positive and negative consequences will thus be thoroughly investigated in this book. The answers offer a much-needed understanding of hope's potential to advance peace.

This book examines these three overarching questions from multiple perspectives and methodologies. Still, readers might rightly wonder whether it makes sense in the first place to study hope in a seemingly hopeless context such as intractable conflicts. Indeed, choosing long-lasting violent conflicts as suitable cases to study hope may appear odd, even absurd. My answers to this legitimate question stem from two standpoints, one intellectual and one normative. From an intellectual, research-oriented standpoint, I find it highly instructive to study hope when it is most challenged. Examining hope in the extreme political circumstances of intractability can teach us about the boundaries of hope and its nuances. Does it make sense to hope for peace after decades of hostility and war? Is it not more appropriate to be hopeless?

A good allegory I often use when I talk and write about hope and conflict is the condition of severe physical illness. We all get sick in the course of our lives, and, in some situations, we, or our loved ones, might develop severe illnesses. People in these difficult situations debate within themselves and with others about the possibility of recovery and what needs to be done, if anything, to get well. Interesting questions about hope arise. What characterizes those who have high hope for recovery even when they face extreme illness? What typifies those who are hopeless even though their sickness is not that severe? Most importantly, what are the consequences of hope in these dire circumstances? Do the hopeful have more chances to survive? If so, why? If not, is it still worthwhile to hope? Intractable conflicts are like incurable illnesses to the extent that both stretch hope to its limits.[2]

The second standpoint is normative and involves the conviction that good social science aims to change human behavior for the greater good. As a conflict scholar and political psychologist, I see it as my mandate and obligation to turn every stone to transform conflict and hasten peace. Luckily, and unlike physical illness, intractable intergroup conflicts are human-made. People create conflicts, and, as proved by dozens of peace agreements signed after long-standing disputes, people are the ones who solve them. As I will show, hope is a prerequisite to conflict resolution and an idea that must be nurtured and cultivated to pursue peace. However, hope has several shortcomings that must be acknowledged and, as much as possible, avoided, if hope is to be used effectively in the transformation of conflict. Investigating the underpinnings and consequences of hope where it is most needed is thus a worthy endeavor from a normative position as well.

It might be necessary to clarify that the ideas and empirical investigations outlined in this book do not seek to make the case that intractable conflicts are solvable by peaceful means. Nor do I suggest that intractable conflicts cannot be peacefully resolved. Estimating whether peace is a probable or improbable outcome of an intractable conflict is not, in any way, the goal of this study. Instead, this book unravels the nuanced interplay between the reality of conflict and the experience of hope (and hopelessness). Do subjective experiences of hope and "objective" realities influence each other? If so, is the relationship

[2] Many illuminating insights about hope are found in the literature on nursing and palliative care that explores patients' hope for recovery when the chance for healing is slim or even absent (e.g., Herth, 2005; Olsman, 2020; Olsman et al., 2015).

reciprocal? More specifically, I examine whether hopelessness contributes to the longevity of conflicts and whether hope can contribute to their resolution. The book proposes exploring hope as a source of conflict transformation and provides initial evidence that this approach is helpful for theoretical analysis and practical implementation.

This introductory chapter continues as follows. I begin by describing the theoretical and disciplinary approaches used in the book, followed by an in-depth look at the main theoretical departure point of the monograph: the theory of the sociopsychological infrastructure of intractable conflicts. Next, I present the empirical bases of the book and summarize the book's key takeaways. I then elaborate on the book's theoretical and applied contribution and continue with some brief notes about terminologies used throughout the book. The introduction chapter concludes with a detailed map of the chapters to come.

Disciplinary and Theoretical Approaches

This book is concerned with the relationship between two concepts: hope and conflict. The former is arguably a subjective experience, whereas the latter is an objective one. Investigating such distinct domains calls for a multidisciplinary approach that connects the dots between epistemes, perspectives, and methodologies. Regrettably, existing research on hope has been crippled by isolated disciplinary approaches. Accordingly, this book aims to reexamine hope by initiating a dialogue between the two main disciplines that explored it: philosophy and psychology.

Philosophers have been discussing hope for hundreds, even thousands of years, providing valuable insights into hope's properties and nature. Philosophy also offers essential debates about the benefits and perils of hope and its role in the history of humankind. Psychologists have been looking into hope since the 1960s, exploring hope from the perspectives of evolutionary, developmental, cognitive, affective, and positive psychology. Using empirical scrutiny, psychology teaches us about the mechanics of hope and its cognitive, affective, and behavioral correlates. The need to merge philosophical and psychological perspectives, the first providing the "soul" and the second the "body" of hope, became evident throughout the years as I advanced in my explorations into this fuzzy and elusive concept. To the best of my knowledge, this book is the first attempt to merge these two perspectives into a coherent story and the first to use them to examine a devastating social condition—intractable violent conflict.

The concept of conflict, specifically intractable conflict, will be explored from the vantage point of political psychology. The impressive theoretical and empirical advances made in the past two decades by political psychologists studying intractability led to a provocative claim; that the most impactful factors contributing to the continuation of intractable conflicts are psychological, namely, subjective interpretations, beliefs, emotions, and attitudes prevalent among those enmeshed in the dispute (Bar-Tal, 2013, 2019; Coleman, 2003; Oren, 2019). The central premise of this approach, termed the "sociopsychological infrastructure of intractable conflicts" (Bar-Tal, 2007, 2019), is that protractedness allows enough time for detrimental perceptions, beliefs, emotions, and attitudes to be cemented and internalized by members of the rival parties. As time passes, the material factors that prompted the conflict are replaced by a set of psychological factors, such as negative perceptions about the outgroup (e.g., their intentions are malevolent), positive perceptions about the ingroup (e.g., our intentions are benevolent), and deterministic perceptions about the conflict (e.g., the conflict is irreconcilable). These and other psychological factors combine to formulate a *sociopsychological infrastructure* that eventually becomes the crux of the conflict. According to the theory, intractable conflicts are perpetuated by this infrastructure, and so their resolution must involve the dismantling of psychological factors embedded in the minds and hearts of members of the rival societies.

To better understand the sociopsychological infrastructure of intractable conflicts, scholars began to examine its cognitive, affective, and behavioral components. Research has focused, for example, on outgroup hate and dehumanization (e.g., Adwan et al., 2016; Halperin et al., 2011), ingroup victimization and entitlement (e.g., Endevelt et al., 2021; Schori-Eyal et al., 2014), and the belief that one's conflict is "unique" (e.g., Kudish et al., 2015). Hope and hopelessness have also gained the attention of political psychologists studying intractability (e.g., Cohen-Chen et al., 2019; Halperin et al., 2008; Hasan-Aslih et al., 2019; Rosler et al., 2017). However, two concerns arise when looking at the existing work on hope and conflict.

First, it appears that most scholarship used an unstandardized way to conceptualize and operationalize hope. For instance, some scholars studied participants' hope for peace as an expression of participants' *desires* for peace, whereas others examined participants' hope for peace as an expression of participants' *belief in the likelihood* of peace. Without a standardized definition, the results were mixed, sometimes contradicting. I elaborate on the challenges arising from how hope was conceptualized and operationalized in past research (see Chapter 3) and then offer a concise model that rectifies the confusion about how to conceptualize and operationalize hope. Second, most research on hope during conflict focuses on psychological mechanisms while failing to ask more substantial questions about hope's role on the domestic and international levels. The result is that both conflict scholars and political psychologists are somewhat lost when it comes to understanding how hope functions in conflict. This book addresses both limitations. To advance the conceptualization and operationalization of hope, I clarify what hope is and how it can be measured. To provide a fuller context about hope's role in conflict, I merge perspectives from philosophy, psychology, political science, international relations, and history.

I have outlined the main disciplinary approaches of the book but not the chief theoretical framework guiding its assumptions. Although I build on several theories, the book's theoretical point of departure is the theory of the sociopsychological infrastructure of intractability (Bar-Tal, 2007, 2019). I now expand on the basic principles of the theory and its use in the analysis of conflict.

Intractable Conflicts and Their Sociopsychological Infrastructure

Among the most severe international conflicts are those defined as "intractable." These long-standing disputes, also termed "protracted conflicts" (Azar, 1990), "deep-rooted conflicts" (Burton, 1987), or "enduring rivalries" (Goertz & Diehl, 1995), are conflicts that last more than a generation, and are marked by hostilities and tensions between the belligerent parties (Bar-Tal, 2013).[3] Intractable conflicts revolve around existential issues, operate on multiple levels, and are total in their detrimental impact on public life and political institutions (Coleman, 2003; Zartman, 2005). To cope with the multiple threats imposed by violence and tensions, people embroiled in intractable conflict establish effective sociopsychological mechanisms in the form of epistemes and narratives shared by members of society (Bar-Tal, 2007). These shared ideas, like the beliefs that the ingroup is benevolent, the outgroup is evil, and the conflict is irreconcilable, feed back into the dynamics of the conflict and promote its continuation (Bar-Tal et al., 2014).

The ongoing conflict between India and Pakistan over Kashmir is an example of an intractable conflict. Since the late 1940s, India, Pakistan, and Kashmiri

[3] The terms "protracted conflict," "enduring rivalries," "deeply rooted conflict," and "intractable conflict" are closely related but not synonyms. For further elaboration of the concepts see Azar, 1990; Goertz and Diehl, 1993; Kelman, 2018; Kriesberg, 1998.

separatists have been battling over the control of Kashmiri land and population. The conflict is marked by a religious component, as Hindus and Muslims, both in Kashmir and in the respective contesting states, clash to gain religious and political hegemony over Kashmir (Behera, 2016; Schaffer & Schaffer, 2005). Overt instances of violence include four full-fledged wars, dozens of border skirmishes, small-scale guerilla warfare, extrajudicial killing, violent riots, and police brutality (Suedfeld & Jhangiani, 2009). Most alarmingly, both countries possess nuclear weapons to leverage their power against the opponent. As in other intractable conflicts, attempts to seek resolution by drawing lines on the map proved futile (Schaffer & Schaffer, 2005). Though the death toll has declined in recent years, the conflict is an ongoing problem yet to be resolved.

The dispute between the Greek and Turkish Cypriots over land, identity, and political power is another example of an intractable conflict (Hadjipavlou, 2007; Heraclides, 2011; Loizides, 2007; Psaltis et al., 2019). Tensions began in the late 1950s, as the British mandate over the island dissolved. In 1974, tensions escalated into violence, including a coup d'état, a military invasion, and the displacement of thousands. Since then, the conflict has been resilient to international and domestic resolution attempts. Since 1974, UN forces have been deployed in the buffer zone between the north and the south to prevent war. The diplomatic deadlock between Greek and Turkish Cypriots and other stakeholders seems insurmountable. Like other intractable conflicts, the dispute has been resilient to international involvement to solve it, including a failed UN-sanctioned referendum for peace that was turned down by most Greek Cypriots.

Perhaps the most intractable of them all is the Israeli-Palestinian case. It has been almost a hundred years since the clashes between Jewish and Palestinian residents of—then-British-ruled—Mandatory Palestine turned into a lethal confrontation between two national movements seeking to establish their homelands between the Jordan River and the Mediterranean Sea (Cohen, 2015). Since then, the Jewish and Palestinian national movements have been fighting over territory, sovereignty, and legitimacy by every means possible. The battle is relentless, uncompromising, and deadly. It involves large-scale wars, thousands of violent confrontations, cruel diplomatic warfare, and the total commitment of citizens to sacrifice their lives to defeat the adversary.

World politics have dramatically changed during this century, which saw two world wars, the rise and fall of Cold War politics, and the demise of Western imperialism. Man has landed on the moon, eradicated polio, and invented the computer, the internet, and Generative AI. The political makeup of the Middle East has also entirely shifted during this era. Nevertheless, the deadly conflict in the Holy Land has persisted throughout these massive changes, serving as an alarming constant in international politics. The international involvement in trying to bring the sides to an agreement is unprecedented, but the conflict seems further from resolution than ever.

The conflicts in Kashmir, Cyprus, and Israel-Palestine are typical examples of intractable conflicts. They all began when groups' goals were understood to be incompatible (Bar-Tal et al., 1989; Mitchell, 1981).[4] The conflicts' inception was thus marked by violent competition over resources, followed by an increase in the conflict's material, human, and psychological price. The theory of the sociopsychological infrastructure of intractability posits that to cope with the complex realities of the conflict, societies form a comprehensive set of beliefs, perceptions, attitudes, and emotions which, once in place, help group members organize their opinions and thoughts about the conflict, provide a rational for the chaotic conditions of conflict and war, and offer a sense of control (Bar-Tal, 2013; Rouhana & Bar-Tal, 1998).

For example, to endure the material and psychological costs of conflict and be willing to kill and sacrifice their children's lives, people must be convinced that their group's goals are just, moral, and rightful (Bar-Tal, 2007; Klar & Baram, 2014). After all, without a sturdy, unshaken belief that the goals are moral and rightful, the group would lose its cohesion and motivation to fight. These ideas are thus promoted by all available means, including the educational system, leaders' rhetoric, and the media (Oren, 2019). Over time, self-justification becomes ingrained in society to the extent that people become blind to their share of responsibility for the conflict's continuation. Another rigid belief held by those living in intractable conflict is that the adversary is immoral and subhuman. Among other negative ramifications, denying the outgroup's humanity helps ingroup members justify past and contemporary atrocities committed against the "enemy" (Bar-Tal et al., 2014; Shulman et al., 2020; Staub, 1990).[5]

During the formative stages of intractable disputes, these and other psychological constructs solidify and become the main prism through which members of each party experience reality. As the conflict progresses, these mental constructs become inseparable from how elites and the public feel, think, and behave. Group members' sense of vulnerability and victimhood, support for armed action, and reluctance to compromise stem, at least in part, from the sociopsychological infrastructure of the conflict. Even in the rare

[4] Whether these goals are indeed incompatible—if there is such a notion as pure incompatibleness in the social world—makes little difference.

[5] It has become a habit of Israeli Parliament Members to ridicule Palestinians and mock their existence as a social group. For example, speaking from the podium of the of the Knesset, MK Oren Hazan (Likud Party) claimed that Palestinians are less worthy than cheap dishwashing detergent and MK Anat Barko (Likud Party) said that there is no such thing as Palestinians because Arabs can't pronounce the letter "P."

cases where international and regional factors align to enable conditions conducive to peace, the rigid sociopsychological infrastructure pulls the societies and their leadership back into antagonistic and hostile prepositions. For this reason, the sociopsychological infrastructure of the conflict serves as one of the biggest obstacles to resolution (Leshem & Halperin, 2022).

To further explain the role of the sociopsychological infrastructure as a perpetuator of intractable conflict, it will be helpful to utilize the dynamical systems approach (Vallacher et al., 2010). One of the most perplexing characteristics of intractability is that it involves two contradicting features: dynamism and stability (Coleman et al., 2007). Intractable conflicts are dynamic because external factors, both international and domestic, change and transform during the conflicts' decades-long duration. International superpowers rise and decline in their influence over the conflict while national leaders and domestic stakeholders who played a pivotal role in the conflict's inception are replaced by other leaders and stakeholders with different constituents and agendas, and then replaced again. At the same time, intractable conflicts seem completely stable. Scenarios of escalation and de-escalation appear in a repetitive pattern (Coleman, 2003). Threats, acts of hostility, stagnation, escalation of overt violence, and failed negotiation attempts become part of a constant, almost foreseeable routine. The moving parts of the conflict seem to be continuously changing, while the conflict system as a whole remains stationary (Kriesberg, 1998). The actors change, but the "show" stays the same.

What element makes these conflicts so stable, even in the face of tectonic shifts in domestic, regional, and international factors? What can serve as the powerful anchor holding the conflict in place? The theory of the sociopsychological infrastructure suggests that this anchor is the rigid infrastructure of beliefs, attitudes, perceptions, and emotions ossified in the "hearts and minds" of citizens and elites (Bar-Tal, 2007). In essence, once parties to a conflict have developed stable ways of feeling, thinking, and behaving toward one another, the problem no longer revolves around the issues in dispute per se (Kelman, 1999). Instead, the conflict is perpetuated by the fixed psychological factors transmitted from one generation to the next. Once anchored, the sociopsychological infrastructure becomes the governing system of the dispute and the dominant force driving intractability.

Given the high price they pay, people mired in the devastating circumstances of intractable conflicts should have the greatest interest in opting for resolution.[6] After all, even a partial solution to the conflict should

[6] Of course, conflict is not necessarily costly for everyone because leaders can profit from the continuation of the conflict in multiple ways (Licklider, 2005). First, focusing on the conflict distracts public's

be preferred over the continuation of enormous costs, anguish, violence, and death. Furthermore, the perpetuation of the conflict over an extended period should provide clear evidence that absolute victory is improbable; ergo, peaceful resolution should be the preferable choice for both parties. However, intractable conflicts persist despite the vast costs the belligerents need to endure (Coleman, 2003, Vallacher et al., 2010). Contrary to reason, parties in intractable conflict seem to prefer a hurting stalemate to resolution and, in some cases, revert to the self-defeating cycle of violence even after a period of collaboration (Goldstein et al., 2001; Pruitt, 1997). The tendency of intractable conflicts to protract at the expense of self-interest further suggests that intractable conflicts are anchored in pervasive, rigid, often subconscious psychological factors (Agnew, 1989; Coleman, 2003). One of these psychological factors is the lack of hope for peace.

The Lack of Hope for Peace in Intractable Conflicts

The wide array of psychological constructs (beliefs, attitudes, perceptions, and emotions) that comprise the sociopsychological infrastructure are directed not only toward the ingroup or the outgroup but also toward the conflict itself. One of the most common beliefs is that the conflict is inherently irresolvable (Rouhana & Bar-Tal, 1998). Many people living in intractable conflict zones believe the conflict will continue forever. To illustrate, as of 2021, approximately two-thirds of Cypriots (55.5% among Turkish Cypriots and 78.5% among Greek Cypriots) believe that the chances of solving the island's intercommunal dispute are meager. Adamant beliefs in the innate irreconcilability of a conflict are perhaps best exemplified in the case of the Palestinian-Israeli conflict. For example, data I collected in 2022 show that 38% of Israelis and 31% of Palestinians (the mode in each society) assessed the chances of an agreement at any point in time as nil. Indeed, decades of hostility and violence, coupled with numerous failed attempts to negotiate an agreement, provide convincing evidence for Israelis and Palestinians that the conflict is innately irreconcilable.

attention from unresolved domestic issues. Because the conflict's continuation is seen only as a result of the adversary's actions (Deutsch et al., 2006), it conveniently exonerates the leader from responsibility for the deterioration of the conflict. Second, political leaders accumulate power based in their ostensible ability to defend the homeland and overwhelm the "enemy" (Zartman, 2005). Ending a conflict will eliminate the need for protection and consequently lessen their power. Third, conflicts bring economic profits to security industries and their sponsors in government. As the conflicts prolongs, these industries become well-established, with ties to leadership. Solving the conflict will hurt a prosperous economy that thrives on war and conflict.

Figure 1.1 Hope for peace: unidirectional model.

The effect of the reality of prolonged intractable conflicts on people's lack of hope for peace can be expressed in a simple figure (Figure 1.1). The nature of the relationship is relatively straightforward. People who live through prolonged intergroup conflict tend to have low hopes for peace. Here, the external reality dictates the belief that the conflict cannot be resolved.

However, as the theory of the sociopsychological infrastructure of intractability suggests, psychological factors are not only the result of the conflict but also the source of it (Bar-Tal, 2013; Coleman et al., 2007; Klar et al., 2013; Vallacher et al., 2010). Can this approach be applied to hope? Does hopelessness facilitate intractability? In his seminal work on track-two diplomacy, the renowned scholar-practitioner Herb Kelman wrote, "Another important element of a supportive environment is the sense of possibility—the sense that, although negotiations may be difficult and risky, it is possible to find a mutually satisfactory solution. This sense of possibility contributes to creating self-fulfilling prophecies in a positive direction, to counteract the negative self-fulfilling prophecies that result from the mutual distrust and pervasive pessimism about finding a way out that normally characterize protracted conflicts" (Kelman, 2010, p. 393).

Kelman's observation can be easily applied outside the diplomatic arena. The lack of hope for peace among the citizenry will likely shape their behaviors in ways detrimental to conflict resolution. Why should people engage in reconciliation or support compromise if they consider the conflict innately irreconcilable? Why should they engross themselves in psychologically taxing tasks like expressing empathy toward the Other or reflecting on the group's wrongdoings if they believe the conflict can never be resolved? There is simply no incentive to work for peace if future peace is perceived as impossible. In the aggregate, a resolution can never be reached when no one is working for peace. Thus, as presented in Figure 1.2, we should expect mutual influence between intractability and hopelessness.

The schematic approach illustrated inFigure 1.2 provides a mere starting point for the book's inquiry into the relations between conflict and hope. It assumes a two-directional pattern of influence and, as such, urges us to explore both paths. More generally, the theory of the sociopsychological infrastructure of intractable conflicts also serves as a convenient departure point

Figure 1.2 Hope for peace: bidirectional model.

for exploring the (lack of) hope for peace during conflict. Starting points are by no means the end of the journey. As will be clearly shown in the remainder of the book, the relationship between hope and intractability is much more nuanced and, in some cases, counterintuitive.

Data on Hope Amidst Conflict

This book provides examples, data, and analyses from diverse sources and contexts to shed light on the complexities of hope amidst conflict. From Greek mythology to modern Marxist thought, from contemporary American politics to the intractable conflicts in Cyprus and Ireland, the book tackles the link between hope and conflict from multiple perspectives. Yet much of the book's insights on hope and hopelessness derive from the conflict that many consider as the prototypical case of intractability, the case of the conflict in Israel-Palestine (Bar-Tal, 2013). More specifically, a substantial part of the book's empirical backbone is based on original data I have been collecting in Israel-Palestine since 2016 and on published and unpublished research I and others have conducted in the region. The original data I present and analyze, some of which are introduced here for the first time, is methodologically diverse and includes correlational, experimental, qualitative, and text-based data on hope and hopelessness in Israel-Palestine. I now briefly elaborate on the original data in the book.

First, the correlational data I present, analyze, and discuss are taken from the Hope Map Project, a large-scale study I conducted on representative samples of Palestinians from the West Bank and the Gaza Strip and Jewish Israelis from Israel proper. The Hope Map Project was designed to capture the levels of hope for various types and definitions of peace and reveal the sociopsychological and political correlates of hope for peace. More specifically, I present the factors that predict hopes for peace among those mired in conflict (Chapter 4) and the political and policy-oriented outcomes of hope (Chapter 7). Another set of data presented in the book is taken from published and unpublished experimental work on hope inducement. Hope-inducing experiments conducted in the context of conflicts seek to test the causal link between hope and peace-supporting attitudes and behaviors. In these studies, my colleagues and I sought to test how hope can be instilled and whether instilled hope can create a positive change in people's conflict-related attitudes and behaviors (see Chapter 7). Will people be more willing to support compromise once their hopes for peace are raised? Will they enroll in peacebuilding programs if we (experimentally) increase their hopes for peace?

Two other sources of original data illuminate how hope functions amidst conflict. The first shifts our attention from society members to their leaders by analyzing how leaders use hope and skepticism in their speeches. To investigate leaders' rhetoric of hope, we analyzed hopeful versus skeptical utterances made over the past twenty-three years by Palestinian and Israeli leaders speaking at the UN general assembly (Chapter 6). The linguistic analysis of this corpus sheds unprecedented light on the connection between hope, power, and the rhetoric of those leading nations entrapped in conflict. The second source of original data comes from what one might call the "opposite end." Here we looked at hope and hopelessness among Israeli and Palestinian grassroots activists working to advance peace from the bottom up (Chapter 5). The study included a thematic analysis of in-depth interviews unraveling the role of hope in the lives of Israeli and Palestinian peacebuilders whose hopes are constantly challenged by intergroup hostilities and war.

Main Takeaways

Three overarching conclusions evolve from the theoretical and empirical examination of hope amidst conflict and constitute the book's main takeaways. Pertaining to the question of what is hope, the first takeaway is that hope is not a one-dimensional construct but a bidimensional construct comprised of two independent dimensions; the *wish* (desire, aspiration) to attain a goal (in our case, "peace") and the expectations (assessments, estimations) of attaining it (Chapter 3). I first show that the dimensions are discrete and, to a great extent, independent. I then demonstrate that different factors influence each dimension and that each dimension has a different role in the hope–conflict dynamics.

Embedded in theory and supported by the data, the *bidimensional model of hope* also addresses some of the limitations the existing literature on hope and conflict. The book further demonstrates that conceptualizing and operationalizing hope as a bidimensional construct creates fruitful avenues for theoretical and applied work. For example, both philosophical deliberation and empirical findings provide initial hints that there is a hierarchy between the two dimensions, such that wishing for an outcome may be more important than believing it can materialize.

Second, one of the most critical takeaways from the theoretical and empirical exploration of the determinants of hope is based on the notion that people's hope can relate in two distinct ways to their present political reality. Depending on the context, people's hopes for the future can be congruent with their political circumstances or in defiance of it. Placing hope in asymmetrical intergroup conflict, the book suggests that disadvantaged group members and their leaders are inclined to adopt a more hopeful stance toward future peace because the dire reality of conflict and suffering commonly experienced by the lower-power party demands hoping for a better future. On the other hand, citizens and elites from the advantaged group can "afford" to be hopeless because they live a more bearable version of the conflict. Chapters 4 and 6 demonstrate and elaborate on the thought-provoking connection between power and hope as well as define its boundaries and limitations.

The third takeaway is that hope can be an effective pathway toward conflict transformation under certain conditions. Whether hope can generate social change has been debated by thinkers and scholars since Spinoza. However, empirical research has only recently tested whether hope catalyzes social change. In this regard, the book is an extensive study within the new literature on hope that looks at the consequences of hope amidst conflict (Chapter 7). Fusing psychological and philosophical insights with new correlational and experimental data, I demonstrate hope's pivotal role in creating the conditions for peace.

However, I also address the dangers and drawbacks of hope, like the understandable concern that hoping for political goals that are extremely difficult to achieve might leave people even more desperate when these hopes are dashed. The main concern is that dashed hopes may generate irreversible frustration, resignation, and setback. Though work on hope outside intergroup conflict demonstrates that being hopeful is advantageous even under challenging circumstances, the book devotes much attention to the downsides of hope, such as misguidance, detachment, and naivete (Chapter 2). Merging the benefits and shortcomings of hope, the book advocates for an optimal type of hope that exploits hope's positive, energizing features while avoiding its dangerous pitfalls in the form of unrealistic prophesizing. Perhaps provocatively, I claim that optimal hope in intractable conflicts should be based on the unquestionable desires to achieve peace, not necessarily the expectations that peace will be achieved.

Contribution

The book contributes to existing literature in at least four ways. First, the contemporary empirical literature on hope for peace fails to connect with earlier philosophical scholarship about hope's role and meaning in the life of individuals and societies and thus lacks a broader structure to be embedded in. *Hope Amidst Conflict* provides this structure by making the necessary links between recent empirical research and the vibrant intellectual debate on hope among philosophers, theologians, and political thinkers. Second, the book addresses the inconsistency in the recent wave of hope research about how hope should be conceptualized and operationalized. This inconsistency resulted in mixed results about the levels of hope among people embroiled in prolonged conflicts. *Hope Amidst Conflict* rectifies this inconsistency by offering the bidimensional model of hope, which provides a clearer understanding of what hope is, what it is not, and how it can be measured. The rationale behind the model is thoroughly described, as are the model's benefits in studying hope in various political contexts.

Third, *Hope Amidst Conflict* offers new insights from the most comprehensive database on hope for peace collected in a conflict zone. The Hope Map Project, conducted simultaneously in Israel, the West Bank, and the Gaza Strip, provides a detailed account of Palestinians' and Israelis' hopes for the future. The study is unique in its scope and ability to compare the attitudes of two groups involved in a high-intensity conflict. Together with the linguistic analyses of Palestinian and Israeli leaders' speeches and the thematic analyses of interviews with Palestinian and Israeli activists, this book offers a diverse set of primary sources shedding new light on how hope functions in conflict.

Last, the book contributes to applied conflict resolution by reintroducing the deliberate use of hope as a tool for political change among leaders and peacebuilders working in conflict zones. For example, the hope-inducing interventions presented in this book serve as blueprints for media campaigns that challenge the detrimental perception that peace is impossible. Furthermore, working in the often-discouraging settings of protracted violent conflicts, activists and community leaders commonly experience fluctuations in their own levels of hope. Finding ways to nurture and maintain hope for peace can be incorporated into capacity-building programs focused on the peacemakers themselves.

Terminologies

Some clarifications about terminologies are due. I start with the "simple" task of clarifying what I talk about when I talk about peace. People often use the word "peace" offhandedly, assuming, perhaps, that everybody knows what peace "is." However, it stands to reason that different people have different interpretations of peace. Some scholars, including myself, have taken an interest in the different interpretations of peace and argued that notions of peace could be clustered around three peace "types" (Biton & Salomon, 2006; Galtung, 1969; Leshem & Halperin, 2020). First, peace can be understood as the absence of war and bloodshed. This type of peace is called "negative peace," as it involves negating overt violence (Boulding, 1977; Galtung, 1969). Second, peace can be interpreted as positive social relationships characterized by amicability, collaboration, and harmony. This interpretation is called "positive peace" since it implies positive intergroup interaction (Hakvoort & Hägglund, 2001; McLernon et al., 1997). The third interpretation refers to peace as an uncoercive sociopolitical system where equality and justice prevail. This type of peace is referred to as "structural peace" as it seeks to overturn social structures that perpetuate systemic violence (Burton, 1997; Rubenstein, 2017).

Studies show that the different interpretations often align with group power (Biton & Salomon, 2006; Leshem & Halperin, 2020). High-power group members tend to think of peace as "positive peace" (i.e., harmony, friendship) rather than "structural peace" (i.e., justice, freedom), whereas members of the low-power group will exhibit an opposed pattern. Research shows, however, that both low- and high-power group members firmly and equally understand peace as "negative peace" (i.e., absence of violence) (Leshem & Halperin, 2020). Due to its relative consensus, I use the word "peace" to refer to a condition that terminates the state of intergroup violence. As I describe in Chapter 3, I provide specific definitions of peace in surveys where participants are asked about their hope for "peace." Providing clear definitions of peace help participants remain on the same page concerning the type of peace they are asked about. Issues concerning the definitions of peace are elaborated in Chapter 4.

Another note on terminology concerns the geographical region of interest. In my writings and lectures, I refer to the geographical area between the Jordan River and the Mediterranean Sea as "Israel-Palestine." I find the term helpful (though perhaps somewhat clumsy) because it is inclusive and acknowledges both groups' connection to the land. I also use internationally accepted terms to distinguish between the geographical areas within Israel-Palestine. I use "Israel proper" when referring to the area west of the Green Line and the "West Bank" to refer to the area east of the Green Line. I refer to the West Bank and Gaza Strip as the Occupied Palestinian Territory (OPT).

Chapter Outlines

The book is divided into eight chapters, each contributing a different perspective on hope and conflict. Following the introduction, Chapter 2 describes the merits and dangers of hope as elucidated by thinkers and social scientists. At first, the reader might find it surprising that hope has downsides and that pessimism can be quite advantageous. In fact, many philosophers were preoccupied with the negative consequences of hope, including Nietzsche, who sharply remarked that "Hope is the worst of all evil for it prolongs that torment of man." Similarly, several cultures and religions regard the expressions of hope as unseemly or a waste of time. Still other thinkers argue that hope is beneficial, even vital, for human survival. Psychological science strengthens the notion that hope is advantageous for individuals and collectives. The second part of the chapter focuses on hope's assets. Overall, the chapter outlines the benefits and disadvantages of hope as manifested in many social and political contexts, including intractable conflicts.

The debate about whether hope is beneficial or detrimental to humankind leads to Chapter 3, which investigates the conceptualization of hope through the two predominant disciplines that study it: philosophy and psychology. The chapter begins with a scholarly dialogue between the two disciplines about what hope "is" and what hope "means." Building on this dialogue, a new model emerges—the bidimensional model of hope—which posits that hope is the amalgam of two discrete components, *wish* and *expectation*. Theoretical and empirical support for the bidimensional model is provided together with examples from everyday life. Hope is then brought back into the political context, with the hypothesis that the intensity of the *wish* to attain a particular political outcome (in our case, "peace") and the *expectation* that this outcome can be attained combine to affect political behavior. The chapter ends by revealing the wishes and expectations for peace of Israelis and Palestinians measured as part of the Hope Map Project.

Chapter 4 answers one of the book's central questions concerning the determinants of hope for peace. The inquiry begins with a presentation of findings on the precursors of hope as revealed in existing studies conducted in conflict zones. I then briefly introduce the methods and key measures of the Hope Map Project and turn to present the demographic, sociopolitical, and psychological antecedences of hope amidst conflict. I first show how age, religiosity, and political ideology are associated with hope for peace. I then show that political efficacy and acceptance of uncertainty predict hope for peace in opposite ways. Last, I demonstrate how Palestinians' and Israelis' perceptions of threat correlate with each dimension of hope. I then tie the results to the asymmetrical nature of the conflict and offer several theoretical and applied implications for the findings.

The quantitative analyses presented in Chapter 4 are complemented by a qualitative study in Chapter 5. The chapter, written in collaboration with Shanny Talmor and Eran Halperin, seeks to understand the role of hope for peace among Israeli and Palestinian peace activists working on intergroup conciliation. The thematic analyses of twenty interviews conducted with peacebuilders from Israel, the West Bank, and the Gaza Strip provides a fresh look at how hope functions on the frontline of peacebuilding. First, perhaps counterintuitively, activists' belief in the possibility of peace (the expectation dimension of hope) is not necessarily high. Rather, it appears that what makes activists "tick" is their uncompromising desires for peace (the wish dimension of hope). Second, echoing the work of political thinkers, the study suggests that hope does not exist without practicing it. For the peacebuilders we interviewed, hope without action seems meaningless. Third, implicit in their words, it appears that their activism is not only intended to achieve peace but also intended to replenish one of humankind's most basic needs—the need to hope.

Chapter 6, written with Ilana Ushomirsky, Emma Paul, and Eran Halperin moves the discussion from the grassroots level to the arena of international politics. Based on Michael Oakeshott's theory of the politics of faith and skepticism, the chapter demonstrates how leaders of nations in conflict use hope and skepticism in their speeches to international audiences. Palestinian leaders, for example, use the rhetoric of hope because hope fuels their struggle for self-determination and garners support from the international community. Israeli leaders, who are less inclined to change the status quo, use the rhetoric of skepticism to indicate their desire to manage rather than transform the conflict. Using linguistic coding of all forty-six speeches delivered by Palestinian and Israeli leaders during the UN General Assembly Annual Meetings, we show how hope and despair are strategically used by leaders of low- and high-power groups locked in an intractable conflict. We then link the findings to the broader questions of hope's role during conflict and peace.

Does hope for peace matter? This is the central question asked (and answered) in Chapter 7. I first present theoretical and empirical research that suggests that hope's palliative nature might, in certain circumstances, translate into passiveness instead of action toward peace. Perhaps ironically, hoping for peace might replace the pursuit of peace, the former being less demanding than the latter. However, most scholarship on hope (the literature on hope for peace being no exception) demonstrates that hope is a motor for political and social change. I then provide evidence from the Hope Map

Project that corroborates this premise by revealing that hope is the most robust predictor of Israelis' and Palestinians' support for collective action, track-two diplomacy, and compromise for peace. The chapter also reveals the causal connection between hope and peace-promoting outcomes, as demonstrated in hope-inducing experiments. The bidimensional model of hope helps determine the relative contribution of wishes and expectations in eliciting these peace-promoting outcomes. Practical implications are then suggested on how hope can be used as a strategy to advance peace.

The book concludes in Chapter 8 with a summary of the book's primary goals and findings. A particular emphasis is given to applied insights that emerge from the comprehensive examination of hope in conflict, insights that can be used by all those working for peace in conflict zones. Drawing together theoretical and applied conclusions, the chapter refers back to the overarching questions raised at the beginning of the book and sets new agendas and empirical paradigms for future research on hope amidst conflict.

References

Adwan, S., Bar-Tal, D., & Wexler, B. E. (2016). Portrayal of the other in Palestinian and Israeli schoolbooks: A comparative study. *Political Psychology*, *37*(2), 201–217. https://doi.org/10.1111/pops.12227

Agnew, J. (1989). Beyond reason: Spatial and temporal sources of ethnic conflict. In L. Kriesberg, T. A. Northrup, & S. J. Thorson (Eds.), *Intractable conflicts and their transformation* (pp. 41–52). Syracuse University Press.

Azar, E. (1990). *The management of protracted social conflict: Theory and cases*. Gower.

Bar-Tal, D. (2007). Sociopsychological foundations of intractable conflicts. *American Behavioral Scientist*, *50*(11), 1430–1453. https://doi.org/10.1177/0002764207302462

Bar-Tal, D. (2013). *Intractable conflicts*. Cambridge University Press.

Bar-Tal, D. (2019). The challenges of social and political psychology in pursuit of peace: Personal account. *Peace and Conflict: Journal of Peace Psychology*, *25*(3), 182–197. https://doi.org/10.1037/pac0000373

Bar-Tal, D., Kruglanski, A. W., & Klar, Y. (1989). Conflict termination: An epistemological analysis of international cases. *Political Psychology*, *10*(2), 233–255. https://doi.org/10.2307/3791646

Bar-Tal, D., Oren, N., & Nets-Zehngut, R. (2014). Sociopsychological analysis of conflict-supporting narratives a general framework. *Journal of Peace Research*, *51*(5), 662–675. https://doi.org/10.1177/0022343314533984

Behera, N. C. (2016). The Kashmir conflict: Multiple fault lines. *Journal of Asian Security and International Affairs*, *3*(1), 41–63. https://doi.org/10.1177/2347797015626045

Biton, Y., & Salomon, G. (2006). Peace in the eyes of Israeli and Palestinian youths: Effects of collective narratives and peace education program. *Journal of Peace Research*, *43*(2), 167–180. https://doi.org/10.1177/0022343306061888

Boulding, K. E. (1977). Twelve friendly quarrels with Johan Galtung. *Journal of Peace Research*, *14*(1), 75–86. https://doi.org/10.1177/002234337701400105

Burton, J. (1987). *Resolving deep-rooted conflict: A handbook*. University Press of America.

Burton, J. (1997). *Violence explained*. Manchester University Press.

Cohen, H. (2015). *Year Zero of the Arab-Israeli conflict 1929*. Brandeis University Press.

Cohen-Chen, S., van Kleef, G. A., Crisp, R. J., & Halperin, E. (2019). Dealing in hope: Does observing hope expressions increase conciliatory attitudes in intergroup conflict? *Journal of Experimental Social Psychology*, *83*, 102–111. https://doi.org/10.1016/j.jesp.2019.04.002

Coleman, P. T. (2003). Characteristics of protracted, intractable conflict: Toward the development of a metaframework-I. *Peace and Conflict: Journal of Peace Psychology*, *9*(1), 1–37.

Coleman, P. T., Vallacher, R. R., Nowak, A., & Bui-Wrzosinska, L. (2007). Intractable conflict as an attractor: A dynamical systems approach to conflict escalation and intractability. *American Behavioral Scientist*, *50*(11), 1454–1475. https://doi.org/10.1177/0002764207302463

Deutsch, M., Coleman, P., & Marcus, E. (Eds.). (2006). *The handbook of conflict analysis and resolution* (2nd ed.). Jossey-Bass.

Endevelt, K., Schori-Eyal, N., & Halperin, E. (2021). Everyone should get the same, but we should get more: Group entitlement and intergroup moral double standard. *Group Processes & Intergroup Relations*, *24*(3), 350–370. https://doi.org/10.1177/1368430219896618

Galtung, J. (1969). Violence, peace, and peace research. *Journal of Peace Research*, *6*(3), 167–191.

Goertz, G., & Diehl, P. F. (1993). Enduring rivalries: Theoretical constructs and empirical patterns. *International Studies Quarterly*, *37*(2), 147–171. https://doi.org/10.2307/2600766

Goertz, G., & Diehl, P. F. (1995). The initiation and termination of enduring rivalries: The impact of political shocks. *American Journal of Political Science*, *39*(1), 30–52. https://doi.org/10.2307/2111756

Goldstein, J. S., Pevehouse, J. C., Gerner, D. J., & Telhami, S. (2001). Reciprocity, triangularity, and cooperation in the Middle East, 1979–1997. *Journal of Conflict Resolution*, *45*(5), 594–620.

Hadjipavlou, M. (2007). The Cyprus conflict: Root causes and implications for peacebuilding. *Journal of Peace Research*, *44*(3), 349–365. https://doi.org/10.1177/0022343307076640

Hakvoort, I., & Hägglund, S. (2001). Concepts of peace and war as described by Dutch and Swedish girls and boys. *Peace and Conflict: Journal of Peace Psychology*, *7*(1), 29–44. https://doi.org/10.1207/S15327949PAC0701_03

Halperin, E., Bar-Tal, D., Nets-Zehngut, R., & Drori, E. (2008). Emotions in conflict: Correlates of fear and hope in the Israeli-Jewish society. *Peace and Conflict: Journal of Peace Psychology*, *14*(3), 233–258. https://doi.org/10.1080/10781910802229157

Halperin, E., Russell, A. G., Dweck, C. S., & Gross, J. J. (2011). Anger, hatred, and the quest for peace: Anger can be constructive in the absence of hatred. *Journal of Conflict Resolution*, *55*(2), 274–291. https://doi.org/10.1177/0022002710383670

Hasan-Aslih, S., Pliskin, R., van Zomeren, M., Halperin, E., & Saguy, T. (2019). A darker side of hope: Harmony-focused hope decreases collective action intentions among the disadvantaged. *Personality and Social Psychology Bulletin*, *45*(2), 209–223. https://doi.org/10.1177/0146167218783190

Havel, V. (1990). *Disturbing the peace: A conversation with Karel Hvížďala*. Vintage.

Hecht, D. (2013). The neural basis of optimism and pessimism. *Experimental Neurobiology*, *22*(3), 173–199. https://doi.org/10.5607/en.2013.22.3.173

Heraclides, A. (2011). The Cyprus Gordian knot: An intractable ethnic conflict. *Nationalism and Ethnic Politics*, *17*(2), 117–139. https://doi.org/10.1080/13537113.2011.575309

Herth, K. (2005). State of the science of hope in nursing practice: Hope, the nurse, and the patient. In J. Eliott (Ed.), *Interdisciplinary perspectives on hope* (pp. 169–212). Nova Science.

Kelman, H. C. (1999). The interdependence of Israeli and Palestinian national identities: The role of the other in existential conflicts. *Journal of Social Issues*, *55*(1), 581–600. https://doi.org/10.1111/0022-4537.00134

Kelman, H. C. (2010). Interactive problem solving: Changing political culture in the pursuit of conflict resolution. *Peace and Conflict: Journal of Peace Psychology*, *16*(4), 389–413. https://doi.org/10.1080/10781919.2010.518124

Kelman, H. C. (2018). *Transforming the Israeli-Palestinian conflict: From mutual negation to reconciliation* (P. Matter & N. Caplan, Eds.). Routledge.

Klar, Y., & Baram, H. (2014). In defence of the in-group historical narrative in an intractable intergroup conflict: An individual-difference perspective. *Political Psychology*, *37*(1), 1–17. https://doi.org/10.1111/pops.12229

Klar, Y., Schori-Eyal, N., & Klar, Y. (2013). The "never again" state of Israel: The emergence of the holocaust as a core feature of Israeli identity and its four incongruent voices. *Journal of Social Issues*, *69*(1), 125–143. https://doi.org/10.1111/josi.12007

Kriesberg, L. (1998). Intractable conflicts. In E. Weiner (Ed.), *The handbook of interethnic coexistence* (pp. 332–342). Continuum.

Kudish, S., Cohen-Chen, S., & Halperin, E. (2015). Increasing support for concession-making in intractable conflicts: The role of conflict uniqueness. *Peace and Conflict: Journal of Peace Psychology*, *21*(2), 248–263. https://doi.org/10.1037/pac0000074

Leshem, O. A., & Halperin, E. (2020). Lay theories of peace and their influence on policy preference during violent conflict. *Proceedings of the National Academy of Sciences*. https://doi.org/10.1073/pnas.2005928117

Leshem, O. A., & Halperin, E. (2022). Societal beliefs, collective emotions, and the Palestinian-Israeli Conflict. In A. Siniver (Ed.), *Routledge companion to the Israeli-Palestinian conflict* (pp. 42–58). Routledge.

Licklider, R. (2005). Comparative studies of long wars. In A. C. Crocker, F. O. Hapson, & P. Aall (Eds.), *Grasping the nettle: Analyzing cases of intractable conflict* (pp. 33–46). United States Institute of Peace.

Loizides, N. G. (2007). Ethnic nationalism and adaptation in Cyprus. *International Studies Perspectives*, *8*(2), 172–189. https://doi.org/10.1111/j.1528-3585.2007.00279.x

McLernon, F., Ferguson, N., & Cairns, E. (1997). Comparison of Northern Irish children's attitudes to war and peace before and after the paramilitary ceasefires. *International Journal of Behavioral Development*, *20*(4), 715–730. https://doi.org/10.1080/016502597385144

Mitchell C. (1981). *The structure of international conflicts*. McMillian.

Olsman, E. (2020). Hope in health care: A synthesis of review studies. In S. C. van den Heuvel (Ed.), *Historical and multidisciplinary perspectives on hope* (pp. 197–214). Springer International. https://doi.org/10.1007/978-3-030-46489-9_11

Olsman, E., Leget, C., Duggleby, W., & Willems, D. (2015). A singing choir: Understanding the dynamics of hope, hopelessness, and despair in palliative care patients. A longitudinal qualitative study. *Palliative & Supportive Care*, *13*(6), 1643–1650. https://doi.org/10.1017/S147895151500019X

Oren, N. (2019). *Israel's national identity: The changing ethos of conflict*. Lynne Rienner.

Pruitt, D. G. (1997). Ripeness theory and the Oslo talks. *International Negotiation*, *2*(2), 237–250. https://doi.org/10.1163/15718069720847960

Psaltis, C., Loizides, N., LaPierre, A., & Stefanovic, D. (2019). Transitional justice and acceptance of cohabitation in Cyprus. *Ethnic and Racial Studies*, 1–20. https://doi.org/10.1080/01419870.2019.1574508

Rosler, N., Cohen-Chen, S., & Halperin, E. (2017). The distinctive effects of empathy and hope in intractable conflicts. *Journal of Conflict Resolution*, *61*(1), 114–139. https://doi.org/10.1177/0022002715569772

Rouhana, N. N., & Bar-Tal, D. (1998). Psychological dynamics of intractable ethnonational conflicts: The Israeli–Palestinian case. *American Psychologist*, *53*(7), 761.

Rubenstein, R. E. (2017). *Resolving structural conflicts: How violent systems can be transformed*. Routledge, Taylor & Francis Group.

Sabatos, C. (2008). Criticism and destiny: Kundera and Havel on the legacy of 1968. *Europe-Asia Studies*, *60*(10), 1827–1845.

Schaffer, H. B., & Schaffer, T. C. (2005). Kashmir: Fifty years of running in place. In A. C. Crocker, F. O. Hapson, & P. Aall (Eds.), *Grasping the nettle: Analyzing cases of intractable conflict* (pp. 295–318). United States Institute of Peace.

Schori-Eyal, N., Halperin, E., & Bar-Tal, D. (2014). Three layers of collective victimhood: Effects of multileveled victimhood on intergroup conflicts in the Israeli–Arab context. *Journal of Applied Social Psychology*, *44*(12), 778–794. https://doi.org/10.1111/jasp.12268

Shulman, D., Halperin, E., Kessler, T., Schori-Eyal, N., & Reifen Tagar, M. (2020). Exposure to analogous harmdoing increases acknowledgment of ingroup transgressions in intergroup conflicts. *Personality and Social Psychology Bulletin*, *46*(12), 1649–1664. https://doi.org/10.1177/0146167220908727

Staub, E. (1990). Moral exclusion, personal goal theory, and extreme destructiveness. *Journal of Social Issues*, *46*(1), 47–64.

Suedfeld, P., & Jhangiani, R. (2009). Cognitive management in an enduring international rivalry: The case of India and Pakistan. *Political Psychology*, *30*(6), 937–951. https://doi.org/10.1111/j.1467-9221.2009.00736.x

Vallacher, R. R., Coleman, P. T., Nowak, A., & Bui-Wrzosinska, L. (2010). Rethinking intractable conflict: The perspective of dynamical systems. *American Psychologist*, *65*(4), 262–278. https://doi.org/10.1037/a0019290

Zartman, W. I. (2005). Analyzing intractability. In A. C. Crocker, F. O. Hapson, & P. Aall (Eds.), *Grasping the nettle: Analyzing cases of intractable conflict* (pp. 47–64). United States Institute of Peace.

2

The Merits and Dangers of Hope

> Wandering hope brings help to many men, but others she tricks.
>
> —**Sophocles**

Greek mythology depicts the enigma of hope in the famous story of Pandora's Box. Seeking to punish the mortal Prometheus for stealing fire from the Gods, Zeus sent Pandora to Earth with a box she was told never to open. As Zeus expected, curious Pandora did not resist the temptation and opened the lid, releasing a swarm of curses and diseases upon human mortals. Envy and revenge for the mind and colic and rheumatism for the body were some of the horrific blows introduced to humanity that day. Realizing what she had done, Pandora promptly closed the lid. Only one thing remained in the box. It was *hope.*

The reason hope was placed in the mystical box is unclear. Was hope the remedy aimed to be dispersed upon humans to assist them in coping with the hardships of life? Or was it one of the diseases, an illness made to prolong humans' suffering by making them believe in the unachievable? Most historians agree that the Ancient Greeks perceived hope as a malady, not a blessing, as they believed challenging fate was a costly and dangerous endeavor (Menninger, 1959; Moltmann, 1968). It might be surprising to the reader that, across epochs, cultures, religions, and geographical locations, hope is not necessarily understood as something good. From the ancient Greeks to postmodern thinkers, many describe hope in negative terms, as something we need to minimize and even avoid.

This chapter aims to explore the good and bad sides of hope. I find this exploration essential since theoretical and empirical research suggest that both aspects need our attention to fully understand the concept of hope. Intuitively, we might find it odd to think of hope in negative terms. However, this is partially because Western and modern worldviews accentuated the merits of hope and deemphasized its downsides (Eliott, 2005). Yet, certain cultures are more reserved regarding the value of hope (Averill et al., 1990; Averill &

Hope Amidst Conflict. Oded Adomi Leshem, Oxford University Press. © Oxford University Press 2024.
DOI: 10.1093/oso/9780197685303.003.0002

Sundararajan, 2005). Recent research in the psychology of emotions adds evidence to this notion by demonstrating that feel-good emotions such as hope and pride can be harmful in certain circumstances (Cohen-Chen et al., 2020).

The complexity of hope emerges when we think more broadly about whether hope facilitates or hinders human survival. As mentioned, the success of the human race cannot only be attributed to humans' hopes to improve their condition by challenging reality but also to their endurance, flexibility, and adaptivity to difficult circumstances. Hope drives people toward outstanding achievements but may also generate rigidness and stubbornness. Theoretical investigations of the benefits and costs of hope, however intriguing, are not the primary goal of this chapter. I mainly wish to discuss the merits and dangers of hope because they are essential for all those working in intractable environments. My guess is that most conflict scholars and peacebuilding practitioners, perhaps most readers of this book, find it evident that hope is something we should always promote and nurture. Yet, my journey into hope convinced me that we must also be familiar with the adverse sides of hope if we seek to transform conflicts such as those in Israel-Palestine or Cyprus. In this chapter, I show that hope is valued differently across cultures and epochs and that hope has costs we must address if we wish to make a real contribution to peace.

This chapter begins by outlining the perils of hope. I use the wisdom of ancient philosophers and postmodern thinkers to shed light on hope's negative sides. Cultural and psychological accounts of the unfavorable consequences of hope are also presented and discussed. I then turn to investigate the great merits and advantages of hope. Theoretical and empirical research from diverse fields provides evidence that hope is beneficial, even essential, for human survival and advancement. I then combine the insights from both approaches to describe an optimal type of hope—a novel approach that maximizes hope's benefits and avoids, as much as possible, its pitfalls.

The Negative Sides of Hope

Hope as Ignorance

One of the most compelling arguments against hope is comparing it to *reason*. Plato, the father of Western philosophy, described hope as something separated from, even antithetical, to reason (Gravlee, 2020).[1] Unlike reason,

[1] Gravlee (2020) also describes Plato's positive account of hope.

hope is an unreliable guide toward the good and, thus, an intellectual fallacy humans must learn to minimize. According to Plato, hopes are easily swayed and involve investments in the future which are highly susceptible to manipulation. Hope is commonly directed at attaining the good, Plato agrees, but hope is bound to miss because it departs from the structured processes of reason. Two thousand years later, it was Descartes who said, "Hope is a disposition of the soul to persuade itself that what it desires will come to pass" (Descartes, 1649/2015, p. 264). Descartes hints at the human tendency to overestimate the likelihood of outcomes we greatly desire. Truth be told, we are all intimately familiar with the tendency to increment our estimation of success in attaining the goals we desire the most. If our desires contaminate our estimation, how can we base our decisions on hope?

Spinoza, one of the most notable critics of hope, contrasts hope with reason with a straightforward argument. Hope (and fear) are linked with doubt, which humankind should eliminate through the quest for knowledge and truth (Blöser, 2020). If humankind fulfills its destiny and devotes itself to seeking wisdom and knowledge, it will rid itself of the unnecessary burden of fear and hope. For Spinoza, reason will lead us to know the certainties of life and consequently free us from our dependency on hope (Spinoza, 1677/2000). Spinoza's ideas resonate with Freud's opinion that people of sophistication and high intellect do not need hope (Freud, 1927; see also Peterson, 2000). Hence, displaying hope is displaying ignorance and weak-mindedness. The human race should replace hope with the pursuit of firm knowledge, hope critics say. Simply put, when we know, hope becomes irrelevant.

Let us place this argument within the context of the century-old conflict in Israel-Palestine. Is it reasonable to claim that peace between the Palestinians and Israel is possible? After decades of hostilities and numerous attempts to reach an agreement, doesn't it make more sense to conclude that peace in Israel-Palestine is improbable? Note that many international conflicts did not reach the state of a comprehensive peace agreement. Some linger with oscillating levels of hostilities, while in others, the conflict was terminated because one party eliminated the other (Diehl & Goertz, 2000; Goertz & Diehl, 1995). The fact is that peace is not an inevitable outcome of conflict, and so future peace in intractable conflicts (like the one in Israel-Palestine) is not something that can be promised.

Most Palestinians and Israelis think the conflict will continue forever (Leshem & Halperin, 2020; Telhami & Kull, 2013). Thus, for most residents of the Holy Land, claiming that peace in Israel-Palestine is possible is more foolish than claiming peace is an impossibility. Yet the lack of hope for peace is also fueled by the skeptical statements of leaders who might be interested in sustaining the conflict. Indeed, the skeptical outlook on Israeli-Palestinian peace has become a trend in the past decade (see Chapter 6 for thorough analyses of the politics of hope and skepticism among Israeli and Palestinian elites). Politicians who speak about the improbability of peace are considered honest, realistic, and mature, while those who argue that there is hope for peace are regarded as extremely naive.

Hope as Arrogance

Another powerful argument against hope comes from Aristotle, who links hope with arrogance (Gravlee, 2020). Aristotle submits that hopeful people portray a favorable but baseless picture of the future and thus lack the courage to face life's uncomfortable truths. Instead of adapting to unpleasant realities, the hopeful place their confidence in a groundless future. Political realists often cite the dialogue between the people of the island of Melos and the armies of Athens as a demonstration of hope's delusional qualities (Alker, 1988). According to the Greek historian Thucydides, the mighty Athenians advised the weak Melian islanders to surrender without a fight and accept Athens' control over their small islet. Placing their hope on the assistance of the gods and the Spartans, the Melians declined the offer. The Athenians replied, "Hope, danger's comforter, may be indulged in by those who have abundant resources . . . but its nature is to be extravagant, and those who go so far as to put their all upon the venture see it in its true colors only when they are ruined." Indeed, due to the Melians' dependency on hope and reluctance to surrender peacefully, a siege was placed on the island. In its aftermath, the Athenians conquered Melos, killed all the men, and enslaved the women and children.

A similar but milder approach resonates with non-Western perspectives that often value the humble acceptance of reality (Chang, 2001). Several studies in cultural psychology shine a light on the relations between hope and humbleness. Chang (1996), for example, compared Caucasian and Asian Americans' optimism toward their future academic achievement and found that, all else equal, Asian Americans tended to be more pessimistic about their chances of success (though no objective reason for their pessimism was present). Another study showed that Canadians were more optimistic than Japanese (Heine & Lehman, 1995). Chang suggested that Asian cultures that accentuate the merits of unpretentiousness favor pessimism over potentially unwarranted and overconfident optimism. In addition, by accepting the hardships of life (such as failure, misfortune, or injustice), individuals gain a better sense of control over the uncertain future (see also Chang et al., 2010; Chang, 1998). Lowering the stakes by lowering the scope of the things one hopes for may help decrease doubt, ambiguity, and uncertainty.

Furthermore, Averill and colleagues (1990) maintain that, like other emotions, hope is socially constructed according to defined rules and norms. During socialization, people learn what emotional responses, both observable and internal, are culturally appropriate. The authors suggest that one of the social rules about hope is the rule of prudence. This socially sanctioned principle advises people to regard hope as inappropriate when the chances of achieving the goal are considered improbable. While some cultures advise caution when it comes to hope, others, like some Western cultures, provide no limit to what one may hope for and perhaps encourage people to hope for the impossible. In these societies, pursuing one's dreams, *no matter the cost*, is often considered a noble and worthy enterprise. Referring to American society, Bellah (2005) writes about the vain and impulsive hope ingrained in what he terms the American Civil Religion. Peterson (2000) adds that the capitalist consumption of material goods is a "socially sanctioned way of satisfying the optimistic forces that organize the entire [American] culture" (p. 52). Capitalist consumerism is tied to hope because it can only thrive on the implicit assumptions that resources will never deplete and the possibilities to exploit them will never end.

In the context of the intractable Israeli-Palestinian conflict, the critique of hope as arrogance is implicit among "conflict mitigation" advocates in Israel. Israeli advocates of conflict mitigation encourage Israelis to accept the reality of the conflict, not challenge it (Goodman, 2018). After this acceptance, a great burden would supposedly be lifted, allowing Israelis and Palestinians to live without unwarranted delusions of peace and instead devote their attention to minimize hostilities where possible. No doubt, accepting the conflict as irresolvable is easier for the advantaged party (in this case Jewish Israelis) but very difficult for the disadvantaged (the Palestinians), for this means they must accept the unfavorable status quo of subordination.

Hope as Anguish

"Hope is the worst of all evil, for it prolongs that torment of man" is Nietzsche's take on hope (1878/1974). Nietzsche's opinion may be extreme, but a closer look reveals much truth. Our hopes cause us aggravation when they do not materialize. In fact, the more we hope, the greater the pain we experience when these hopes are dashed. Skepticism, on the other hand, protects people from frustration and agony and, as such, is widespread in challenging times and harsh conditions. Continuing this line of thought, people locked in decades of conflict might be hopeless about the chances for peace not only because they perceive the conflict as irreconcilable but also because this pessimistic outlook protects them from the torment of frustrated hope. Why suffer the painful stab of disappointment and grief if one can easily avoid them by being skeptical?

In this context, it might be most appropriate to mention the shattered promise of the Oslo Accords. Many international players and residents of the region pinned their hopes on the Oslo talks of the mid-1990s. The Accords were widely believed to be the most concrete and promising steps toward a just and sustainable solution to the Israeli-Palestinian predicament (Bar-Siman-Tov & Kacowicz, 2014). The three signatories of the Oslo Accords even got the Nobel Prize for Peace for their undeniable courage, and, for a little while, a real sense of euphoric optimism was in the air. However, the Accords collapsed and the promise was gone. It is safe to say that the dashed hopes for peace of Palestinians and Israelis in the years that followed pushed the two societies into a much deeper state of despair. The same can be said of the lack of hope for peace among Cypriots after the 2004 Annan Referendum fell through (Amaral, 2019).

Hope scholars and researchers of conflict and politics should not underestimate the protective powers of skepticism. A skeptical outlook is a mighty shield for those enmeshed in extreme political circumstances. It helps them cope with hardships because it diminishes disappointment and frustration so common in contentious political realities. Those advocating for peace in Israel-Palestine must consider the high costs of dashed hopes. Can another attempt to reach an agreement unintentionally leave people even more desperate when their hopes for peace are crushed again in another futile peace process?

Psychological Drawbacks

Ignorance, arrogance, and anguish are potential downsides of hope. But what about direct psychological shortcomings? Does hope have unique psychological disadvantages? Perhaps one of the most fundamental psychological drawbacks of hope is that it is mentally demanding (Bar-Tal, 2001; Jarymowicz & Bar-Tal, 2006). Hope requires deliberation, creativity, pathway thinking, and complex mental processing, which are all cognitively effortful

(Snyder, 1994). Neurologically, skepticism is tied to avoidance and passiveness, whereas optimism is associated with effort and approach (Hecht, 2013; Tops et al., 2017). So, even on the neuronal level, hope is more costly than skepticism.

The second psychological cost is that hope always entails uncertainty, which most people try to avoid (Fiske, 2010). At least to some extent, humans seek predictability and certainty. The tendency to minimize uncertainty might make people shy away from hoping to change a given situation even when that situation is harmful. The heavy price of uncertainty that comes with hope is manifested in people's inclination to justify the status quo even when the status quo is detrimental (Thórisdóttir & Jost, 2011). Often, the more people are familiar with a harmful situation, the greater their justification of the status quo and the less their certainty about what it is like to live outside it. Imagine people who have been living amidst conflicts for generations. They were born into the conflict, and so were their parents and maybe grandparents. They became accustomed to the conflict through narratives conveyed at home, in textbooks, and in the media. In many ways, the conflict has been one of the most stable sociopolitical factors throughout their life. After decades of conflict, the adverse reality of the conflict becomes a "comfort zone" (Bar-Tal, 2019). Tragically, perceiving the conflict as a stable situation extending into the far future provides a sense of certainty and predictability (Fiske, 2010; Thórisdóttir & Jost, 2011). In short, accepting the dire but familiar reality of conflict can be mentally easier than hoping for something unfamiliar like peace.

Last, hope is costly because of the social expectation that the hopeful must act in pursuit of their hopes. From the Marxist thinker Ernest Bloch (1959) to Barak Obama's "Audacity to Hope" (2006), hope has been framed as an action-oriented idea. Eric Fromm, one of the most notable philosophers of hope, stressed that action must follow hope (Fromm, 1968). Many prominent thinkers and political figures further advocated that hope must be active to be called hope. Rick Snyder's oft-cited *hope theory* (2005) implies that there is no such thing as passive hope and that hope should always be judged with respect to intention and commitment. Indeed, people frequently frown on those who say they hope for something but seem to do nothing about it (Averill et al., 1990). Hoping thus becomes a task that necessitates time and commitment. One way to eliminate this taxing aspect of hope is to be skeptical. The skeptics are not expected to try and are, therefore, exempt from social scrutiny and criticism. Skeptics' freedom from obligation is not only granted by society but also by the skeptics themselves. They may insist, for example, that they have no responsibility in hastening peace because they do not believe it can be done. Adopting a skeptical stance is sometimes the shortest way to be relieved from responsibility.

In sum, perhaps counterintuitively, being hopeful about the prospects of peace might be pretty disadvantageous for people engulfed in protracted disputes. Hope increases the resource-consuming demand to be committed to peace and raises the stakes for disappointment when hopes for peace are dashed. In addition, when the reality of the conflict has been the only reality for decades, hoping for future peace creates an unpleasant sense of uncertainty and unpredictability. We have also seen that hope is, to some extent, antithetical to reason; that it might be unwarranted, inappropriate, and arrogant; and that hope often comes with frustration and pain. Given the strong evidence against peace, the reader could easily conclude that skepticism about peace should be endorsed in intractable conflict.

Nevertheless, one point about skepticism conveys a striking warning that calls for our attention. As noted by many (Kelman, 2018; Levine, 1977; Pomper, 1990; Seligman, 1990; Simkin et al., 1983), pessimism is a self-fulfilling prophecy. For instance, during conflict, pessimism feeds back into the reality of the dispute by crippling citizens' motivation to opt for peace. When no attempt is made in the direction of resolution, the conflict persists or even exacerbates, which, in turn, validates the sense of skepticism. Thus, at least theoretically, a pessimistic outlook should not be regarded only as a mere product of an ongoing dispute but also as one of the drivers of conflict (Kelman, 2010; Pruitt, 1997). It follows that if we wish to resolve conflict, we must seek hope.

The Positive Sides of Hope

At this point, the reader may have lost hope in hope. Evidence has accumulated against it, or at least questioned the common assumption that hope is purely a "good thing." If this is the case, even partly, I have achieved my goal of convincing the reader that hope should be approached carefully. After establishing that hope also has its downsides, it is time to look at the benefits of hope, which seem to be more intuitive. Historically, hope's virtues and merits have been acknowledged in the scriptures of the Old and New Testaments and later by theologians and religious thinkers who placed hope as one of the highest virtues that people can aspire for. Centuries later, many modern thinkers and critical theorists elaborated on hope's advantages, concluding that hope is a beneficial, even essential, factor in the life of humans and society. Psychologists began to examine hope only in the second half of the 20th century, with researchers like Martin Seligman and Rick Snyder leading much of the work on hope as a psychological construct. Almost without exceptions, hope psychologists provide empirical evidence that hope is highly advantageous and has significant contributions to personal well-being and success (see Cheavens et al., 2005; Snyder, 2000). As with hope's shortcomings, hope's assets can also be divided into several categories.

Hope as a Virtue

Hope is considered virtuous in the scriptures of monotheistic faiths. Unlike religions that emphasize circularity, repetition, and reincarnation, monotheistic religions are "historical religions" that perceive time linearly (Brunner, 1956). In Christianity, Islam, and Judaism, history progresses forward in the direction of God's will. Therefore, having hope means having faith in God's ability to steer events in the right direction. It follows that having no hope is expressing doubt in God's power. It is thus no wonder that the word "hope" appears seventy-five times in the Old Testament and fifty-eight times in the New Testament. The Testaments presume that one's hopefulness derives from God and, as such, is an unquestionably desired feature (Eliott, 2005), a virtue to be endorsed to demonstrate one's faithfulness. Similarly, in the Quran, the Prophet Abraham says, "Who can lose hope in the mercy of his Lord except those who have lost the straight path?"

One of the most notable expressions in Judaism of unequivocal hope is the famous saying by Rabbi Nachman from Breslov: "There is no despair in the world whatsoever." Quite plainly, Rabbi Nachman urges us to remember that, even in the darkest times, we must retain uncompromising hope in the Almighty, in His ability to redeem, correct, heal, and transform.[2] According to Fackenheim (1970), hope is a Jewish duty. Indeed, hoping for the Messiah and "the next year in Jerusalem" can almost be considered mandatory practices in Jewish tradition. Hope as a Christian virtue was explicitly advocated by the influential Italian priest and theologian St. Thomas Aquinas (1225–1274), who argued that because hope was a virtue that inevitably involved God, every hope, big or small, should be assumed to rely on God's ability to fulfill it. Thus, being hopeless demonstrates one is heathen, while being hopeful proves one is a devout believer (Nunn, 2005).

[2] "There is no despair in the world whatsoever" has been graffitied by Rabbi Nachman's followers on hundreds of walls across Israel and can be seen on thousands of bumper stickers throughout the country accompanied by a smiley face.

Equating hope with virtue is not only confined to religious faith. As mentioned, one of the features of the modern Western worldview is the belief that people hold their future in their own hands and that "anything is possible" if one believes it can be done. In a way, the unquestioning belief in God's omnipotence was replaced with the belief that humans can accomplish anything if they want it bad enough. Today, at least in the West, we admire those who, against all odds, achieved goals that were presumed impossible. We even hold high esteem for people who relentlessly try to accomplish their goals, even if they fail. One is expected to hope and hope big, both in terms of the levels of enthusiasm and the scope of the hoped-for goal. In short, among both religious and secular, it is considered desirable to be hopeful. It follows that, by and large, being hopeful is the norm, not a character defect. And so, one of the advantages of hope, though somewhat hidden, is that it is socially desirable.

Hope as an Existential Need

In his famous *The Critique of Pure Reason*, Kant (1781) articulated the three pillars of reason. The first is epistemic: "What can I know?" and the second normative: "What ought I do?" According to Kant, the third pillar of reason is "What may I hope for?" The fact that Kant placed hope as a pillar of reason and, consequently, as a pillar of being human might seem surprising. But for Kant and many other thinkers (e.g., Fromm, 1968; Hume, 1740/2003; Tillich, 1965), hope is not an ornament that decorates life, an auxiliary that humankind can do without, or a luxury saved for the fortunate, but an indispensable element of human life. Indeed, since *Homo sapiens* began to gather and hunt, human decisions have included some deliberation of the future (Tiger, 1989). Reproducing, raising offspring, planting crops, conducting transactions, preparing for war, and negotiating peace all include some projection into the future. The evolutionary anthropologist Lionel Tiger argues that a slight bias for overestimating positive outcomes, like the chances to mate and reproduce, has evolved throughout history to enhance human survival (1989, 1999). Without hope, humans may not have dared to explore remote territories, new methods of hunting, and creative ways of survival.[3]

Hoping that something we greatly need or desire will actualize in the future often requires sacrifices in the present. Overall, the capacity to sacrifice immediate and maybe less important things for a greater goal to be potentially

[3] Tiger wrote about optimism, which has many similarities with hope. The differences and similarities between hope and optimism are discussed in Chapter Three.

fulfilled in the future has been essential for human survival and progress in the long run. As Tiger puts it, "A hopeful sense of promise of the future may be as important to communities' welfare as yeast to the rising bread" (Tiger, 1989, p. 15). A series of studies on hope and its correlates support the assumed connection between success and hope. People who score high on scales measuring dispositional hope tend to perform better in cognitive tasks and have improved problem-solving abilities (Chang, 1998; Snyder et al., 1999), exhibit better academic and athletic achievements (Curry et al., 1997; Snyder et al., 1991), have higher perseverance (Peterson, 2000), and better physical and psychological health (see Cheavens et al., 2005, for a comprehensive review).

In addition, past research has found that hopefulness is associated with a decrease in women's likelihood of falling victim to cervical cancer and higher recovery rates from illnesses (Cousins, 1976; Schmale & Iker, 1971). A more recent longitudinal study conducted in the United States may be the most substantial evidence that links hope with survival (Lee et al., 2019). The study, which used data collected over fifty years among more than 70,000 participants, showed that middle-aged men and women who scored high on various trait-hope scales had a higher chance of living after the age of 85. This astonishing finding held true even after accounting for various demographic measures, health conditions, and behavioral factors.

Studies in developmental psychology support the claim that hope is essential for the individual by showing that hope constitutes one of the first qualities to develop in an infant (Eliot, 2005; Erikson, 1964). Erikson was explicit in his assertion that, without hope, further developmental processes are compromised and that hope is "the basic ingredient of all effective as well as ethical human actions" (Erikson, 1964, p. 231). Indeed, one can hardly imagine what it is like to live without any hope whatsoever. Looking at the mental states of death row inmates awaiting their execution exemplifies the horror of complete hopelessness (Smith, 2007). Many of these prisoners develop the Death Row syndrome, leading them to volunteer for execution or commit suicide (Cooper, 2009; Smith, 2007).

Another example of the existential aspects of hope is present in the writings of Viktor Frankl. Frankl is known for his book *Man's Search for Meaning* (1946/2008), which recounts his dreadful experience in the Nazi concentration camps and his quest to find hope and meaning even in dire situations. However, Frankl's encounter with hope and hopelessness, meaning and meaningless dates back to his early psychotherapy work on suicide prevention among adolescents in Vienna. His work with hopeless youngsters contributed to his theory that the central force governing human behavior was the need for meaning. Later, in *Man's Search for Meaning*, Frankl explicitly asserts a causal link between hope, meaning, and human survival. According to Frankl, constantly hoping is a necessity, a vital resource that anyone must retain regardless of the present dire circumstances and probable fatal outcomes. The idea that hope is an indispensable coping mechanism amidst adverse conditions is also stressed in the work of psychologists like Shlomo Breznitz (1986) and Richard Lazarus (1999). Michael Scheier and Charles Carver (1985) as well as other psychologists showed that hope is associated with optimal coping mechanisms with various adversities ranging from fertility problems (Litt et al., 1992) to missile attacks (Zeidner & Allen, 1992; see also Peterson, 2000; Carver & Scheier, 2005).

It seems that hoping is essential to humans in its own right, regardless of whether the hoped-for goal is attainable. We have the "right" to hope for goals, even if their obtainability is questionable. Hope as a right was advocated by John Stuart Mill (1875), who claimed that it is appropriate to hope even when there is no evidence that the hoped-for goal can be attained. Hoping is a healthy practice, beneficial in itself, and so, in some cases, the actual attainment is often of secondary importance (Mill, 1875). The notion that hope has direct benefits regardless of the outcome is exemplified in a study about the joy of buying lottery tickets (Burger et al., 2020). It seems that the mere purchase of a ticket elicits joy among lottery players, independent of the draw results. The researchers suggest that anticipation generates pleasure, strengthening the argument that hope is desirable, independent of results. The idea that hope can exist regardless of a belief in attainment will be thoroughly examined in Chapter 5, where we explore hope among peace activists.

Hope as the Impetus of Human Advancement

The notion that hope is essential and central to human life has been passionately promoted since the Age of Enlightenment (Brunner, 1956). One of the foundations of Enlightenment and late-modern thought is that the historical movement from past to future progresses from lower to higher levels of human capabilities and achievements. Through science, knowledge, and technology, humankind is destined to better its conditions, and so hope as an idea is worthy and warranted (Capps, 1968; Eliott, 2005; Moltmann, 1968; Pinker, 2011). For many of these thinkers, human achievements were made possible only because humans believed in their capabilities to bring them about.

The Algerian-born philosopher Jacque Derrida, for example, advocated that people must work in this life as if the world can be transformed (Russell, 2001). If people do not adopt this approach to the fullest extent—that is, if people are skeptics—life will stagnate. Ernst Bloch might be one of the most notable thinkers who wrote on hope as the catalyst for human advancement. In his three-volume manifesto *The Principle of Hope* (1959), Bloch describes how hope propelled political and social advancement. Several years later, in his lecture "Man as Possibility," he said: "Reality has no fixed size. The world is not finished. It is possible to face the world in a manner that goes beyond mere gripping. . . . 'Accepting things as they are' is not an empirically valid formula. It is no positivism, but, instead, a formula for vulgarity, cowardice, and eventually poverty" (Bloch, 1970, pp. 52–53).

Surprisingly, another advocate of hope is Hobbs (Blöser, 2020). Hobbs is commonly associated with the notion that competition and fear are the originators of all political interactions. According to Hobbs, the natural condition of humans is a constant state of war of all against all, where man's competition over resources determines all other political decisions and actions. It is, therefore, quite surprising that hope emerges in Hobbs's writing as a fundamental factor thrusting humanity forward. Hobbs notes that through humans' appetite for well-being—in short, through their hopes for a better life—societies progress and develop. He then claims that hope is a prerequisite for peace between people and states. The resulting peace is thus a direct consequence of hope.

The central theme that emerges from the above work is that hope is a notion that humanity can benefit from and an essential component for our survival as individuals and collectives. Hope creates opportunities and facilitates advancement, while hopelessness and skepticism create stagnation on both the personal and societal levels. In terms of behavior, hope induces us to act, try, and maybe succeed in improving our lives. I have also outlined that hope can have positive consequences even when the hoped-for goal is not attained. The overall conclusion of many theoreticians and empiricists is that hope is good and moral in its own right. Perhaps hope is one of the features that separate humankind from other species. Animals are confined to act based on the possible, while humans can dream and act on the impossible.

Nevertheless, we must also remember the dangers of hope mentioned in the first part of this chapter. They are many, and they are profound. I have mentioned how hope can distort our assessment of reality and consequently misguide our actions and deeds. I also noted that dashed hopes could hurt and create deeper frustration and passivity. Proponents of hope must also acknowledge that the human ability to adapt rather than change is vital for survival (Hecht, 2013). History shows that people's hopes for social changes can overcome great powers but that sometimes these revolutionary hopes are effectively crushed.

How can we exploit the benefits of hope and avoid its pitfalls? How do we keep our enthusiasm about the possibility of change while ensuring our enthusiasm does not lead us to make unfounded decisions? At the same time, we must ensure that the low likelihood of achieving change does not discourage us completely. Indeed, some of the most significant political achievements in history were made against all odds. Insights from existing work and my observations and empirical research on hope have made me confident that there is a more accurate, appropriate, and perhaps effective way to hope amidst conflict and political unrest.

Optimal Hope

A thirty-year-old man lies in a hospital room waiting for the doctor's prognosis. At the other side of the door, the doctor stalls to think about how to deliver the news. After three months of treatment, it seems that the patient's health has not improved. In fact, the treatment has been entirely ineffective and the chances for recovery have diminished considerably. The way hope is handled here by both patient and physician is tricky and has profound implications.

The doctor decides to describe the situation as it is, not sugarcoat it. She shows the results of the tests and interprets the charts and numbers. She explains that the treatment given to him has been unhelpful. Still, in her voice there is no sign of discouragement or grief. On the contrary, her face is reassuring. She is also filled with a passion for trying another type of treatment. This treatment might not result in complete recovery, she admits. The statistics are against it. However, partial recovery should be seen as a very desirable outcome. Partial recovery will entail some discomfort, and a wheelchair might be necessary in the near future. Yet a wheelchair is not the end of the world, the doctor says. There are still many things to see and do in this life. The patient listens carefully. He is startled, disappointed, and confused. He wants to understand. He seeks more information about the current situation and asks many questions about the alternative treatment the doctor suggested. The chances for complete recovery, maybe any recovery, are slim. He needs to accept this. A wheelchair will be horrible. But perhaps he will learn to live with it. There is so much to live for and so many things to do.

This made-up scenario depicts what might be called *optimal hope.* Before delving into optimal hope, I examine two extreme versions of hope likely to emerge in these circumstances. The first type of hope may be termed *illusional hope*, because it is built on unfounded predictions. This type of hope allows the strong desires (in this case, for recovery) to skew the assessments of possibility. Incrementing our estimation of possibility to match our desire is mostly done unconsciously and unintentionally. It is very natural and understandable to boost expectations for something we greatly wish for, especially in extreme situations of physical illness or social conflict.

Nonetheless, wishful thinking does not make a sound foundation for strategy-making and action. Breznitz (1986), who studied stress and its relations with hope and hopelessness, explains that people in stressful situations may approach adversity with denial. They use denial to cope with unpleasant facts and consequently avoid or discount information that creates an unfavorable picture of the future. Illusional hope can also be found in the context of protracted intergroup conflicts. Indeed, some people living in protracted conflict allow their unshakeable desires for peace to distort their perception of reality by boosting their estimation of the feasibility of peace.

Another type of extreme reaction to difficult situations lies at the opposite end of illusional hope. This type of hope allows our assessments of the low chances for success to diminish our desire for it. In the example above, the patient could have easily succumbed to hopelessness by letting the bad news influence his desire to live. This type of reaction is undoubtedly relevant to hope amidst conflict. The low belief in the feasibility of peace among those mired in intractable disputes might reduce their longing for peace. After all, what good can come out of aspiring for something believed to be unattainable? This approach, in turn, creates passivity and cynicism and contributes to the conflict's longevity. In this case, the lack of hope becomes a self-fulfilling prophecy.

Even in less acute situations, people often find it inappropriate to aspire for goals that are hard to achieve. This, of course, might make sense on the immediate level. However, as described in this chapter, in various contexts, making decisions made only on "cold" calculations of probability is a mistake in the long run. Recall Vaclav Havel's provocative remark that hope has nothing to do with prognosis (Chapter 1). Havel—then a persecuted political dissent, later the leader of free Czechoslovakia—meant that his desires, aspirations, and passions for political change were unrelated to his estimations that this change might actualize. As Havel admits, his hope was propelled more by his unshaken desires to free Czechoslovakia from Soviet control rather than his belief that it could be done.

In the medical scenario described above, the doctor and the patient avoided these two extreme approaches by adopting a different approach to hope. This approach optimizes hope by maximizing its benefits and minimizing its drawbacks.

The central premise of optimal hope is that our desires and expectations should be considered separately lest they influence one another in harmful ways. In extreme conditions (like malign diseases or intractable conflict), optimal hope means that we can decrease our expectations but keep our hopes propelled by our desires. When chances for success are low, a person that optimizes hope is not discouraged by the low probability of success but, fueled with firm and unshakable desires, invests all efforts in bringing about the hoped-for change (see Peterson, 2000). In other words, optimal hope is founded more on the hopeful person's desires, wishes, and aspirations rather than on perceptions of a high likelihood of achievement.

Optimal hope addresses hope's shortcomings described earlier. Optimal hope is not ignorance because it is not guided by overestimating favorable outcomes and seeks, rather than avoids, the search for truth. Optimal hope is not arrogant either, as it accepts reality as "it is" and accounts for the costs of pursuing the hoped-for goal. Maybe most importantly, optimal hope is more immune to anguish because it is not driven exclusively by the belief in fulfillment. Because wants, needs, wishes, and aspirations are the main forces driving optimal hope rather than the high estimations of attainment, disappointment is not as painful and should not cause the hopeful person to give up. It should be admitted that the scope of optimal hope might be more restricted than the scope of hope based on high expectations. In other words, the hoped-for goals of optimal hope may be more modest, at least incremental. If one wants to stay realistic, one might prefer advancing in small steps. Nevertheless, precisely because of its modesty, incremental nature, and cautiousness, optimal hope is more sustainable in the face of ongoing hardships. The book's concluding chapter describes some ideas about the practice and implementation of optimal hope during conflict.

I have reviewed both positive and negative aspects of hope and discussed the implications of these aspects in the context of intractable conflicts. However, I have not defined hope nor discussed how it could be measured. This is the aim of the next chapter.

References

Alker, H. R. (1988). The dialectical logic of Thucydides' Melian dialogue. *American Political Science Review, 82*(3), 805–820. doi:10.2307/1962492

Amaral, J. (2019). *Making peace with referendums: Cyprus and Northern Ireland.* Syracuse University Press.

Averill, J., Catlin, G., & Chon, K. K. (1990). *Rules of hope.* Springer-Verlag.

Averill, J., & Sundararajan, L. (2005). Hope as a rhetoric: Cultural narratives of wishing and coping. In J. Eliott (Ed.), *Interdisciplinary perspectives on hope* (pp. 133–166). Nova Science.

Bar-Siman-Tov, Y., & Kacowicz, A. M. (2014). *Justice and peace in the Israeli-Palestinian conflict*. Routledge.

Bar-Tal, D. (2001). Why does fear override hope in societies engulfed by intractable conflict, as it does in the Israeli society? *Political Psychology*, *22*(3), 601–627.

Bar-Tal, D. (2019). The challenges of social and political psychology in pursuit of peace: Personal account. *Peace and Conflict: Journal of Peace Psychology*, *25*(3), 182–197. https://doi.org/10.1037/pac0000373

Bellah, R. N. (2005). Civil religion in America. *Daedalus*, *134*(4), 40–55. https://doi.org/10.1162/001152605774431464

Bloch, E. (1959). *The principle of hope* (vol. 1–3). MIT Press.

Bloch, E. (1970). Man as possibility. In W. Capps (Ed.), *The future of hope* (pp. 50–67). Fortress.

Blöser, C. (2020). Enlightenment views of hope. In S. C. van den Heuvel (Ed.), *Historical and multidisciplinary perspectives on hope* (pp. 61–76). Springer International. https://doi.org/10.1007/978-3-030-46489-9_4

Breznitz, S. (1986). The effect of hope on coping with stress. In M. H. Appley (Ed.), *Dynamics of stress* (pp. 295–306). Plenum.

Brunner, H. E. (1956). *Faith, hope and love*. Westminister Press.

Burger, M. J., Hendriks, M., Pleeging, E., & van Ours, J. C. (2020). The joy of lottery play: Evidence from a field experiment. *Experimental Economics*, *23*(4), 1235–1256. https://doi.org/10.1007/s10683-020-09649-9

Capps, W. (1968). The hope tendency. *Cross Currents*, *18*(3), 257–272.

Carver, C., & Scheier, M. (2005). Optimism. In C. R. Snyder & S. Lopez (Eds.), *Handbook of positive psychology* (pp. 231–243). Oxford University Press.

Chang, E. C. (1996). Evidence for the cultural specificity of pessimism in Asians vs. Caucasians: A test of a general negativity hypothesis. *Personality and Individual Differences*, *21*(5), 819–822.

Chang, E. C. (1998). Hope, problem-solving, and coping in a college student population: Some implications for theory and practice. *Journal of Clinical Psychology*, 953–962.

Chang, E. C. (Ed.). (2001). *Optimism & pessimism: Implications for theory, research, and practice* (1st ed.). American Psychological Association.

Chang, E. C., Sanna, L. J., Kim, J. M., & Srivastava, K. (2010). Optimistic and pessimistic bias in European Americans and Asian Americans: A preliminary look at distinguishing between predictions for physical and psychological health outcomes. *Journal of Cross-Cultural Psychology*, *41*(3), 465–470. https://doi.org/10.1177/0022022109359691

Cheavens, J., Michael, S. T., & Snyder, C. R. (2005). *The correlates of hope: Psychological and physiological benefits* (J. Eliott, Ed.; pp. 119–132). Nova Science.

Cohen-Chen, S., Pliskin, R., & Goldenberg, A. (2020). Feel good or do good? A valence-function framework for understanding emotions. *Current Directions in Psychological Science*, *29*(4), 388–393. https://doi.org/10.1177/0963721420924770

Cooper, P. (2009). Competency of death row inmates to waive the right to appeal: A proposal to scrutinize the motivations of death row volunteers and to consider the impact of death row syndrome in determining competency. *Developments in Mental Health Law*, *28*(2), 105–126.

Cousins, N. (1976). Anatomy of an illness (as perceived by the patient). *New England Journal of Medicine*. http://www.nejm.org/doi/full/10.1056/NEJM197612232952605

Curry, L. A., Snyder, C. R., Cook, D. L., Ruby, B. C., & Rehm, M. (1997). Role of hope in academic and sport achievement. *Journal of Personality and Social Psychology*, *73*, 1257–1267.

Descartes, R. (1649/2015). *The passions of the soul and other late philosophical writings*. Oxford University Press.

Diehl, P. F., & Goertz, G. (2000). *War and peace in international rivalry*. University of Michigan Press.

Eliott, J. (2005). What have we done with hope? A brief history. In J. Eliott (Ed.), *Interdisciplinary perspectives on hope* (pp. 3–46). Nova Science.

Erikson, E. H. (1964). *Insight and responsibility: Lectures on the ethical implications of psychoanalytic insight*. Norton.

Fackenheim, E. (1970). The commandment to hope: A response to contemporary Jewish experience. In W. Capps (Ed.), *The future of hope* (pp. 68–91). Fortress.

Fiske, S. (2010). *Social beings* (2nd ed.). Wiley.

Frankl, V. (1946/2008). *Man's search for meaning*. Rider.

Freud, S. (1927). *The future of an illusion*. Hogarth.

Fromm, E. (1968). *The revolution of hope*. Harper & Row.

Goertz, G., & Diehl, P. F. (1995). The initiation and termination of enduring rivalries: The impact of political shocks. *American Journal of Political Science, 39*(1), 30–52. https://doi.org/10.2307/2111756

Goodman, M. (2018). *Catch-67: The left, the right, and the legacy of the six-day war*. Yale University Press.

Gravlee, G. S. (2020). Hope in ancient Greek philosophy. In S. C. van den Heuvel (Ed.), *Historical and multidisciplinary perspectives on hope* (pp. 3–23). Springer International. https://doi.org/10.1007/978-3-030-46489-9_1

Hecht, D. (2013). Neural Basis of Optimism and Pessimism. *Experimental Neurobiology, 22*(3), 173–199. https://doi.org/10.5607/en.2013.22.3.173

Heine, S. J., & Lehman, D. R. (1995). Cultural variation in unrealistic optimism: Does the West feel more vulnerable than the East? *Journal of Personality and Social Psychology, 68*(4), 595–607.

Hume, D. (1740/2003). *A treatise of human nature*. Courier.

Jarymowicz, M., & Bar-Tal, D. (2006). The dominance of fear over hope in the life of individuals and collectives. *European Journal of Social Psychology, 36*(3), 367–392. https://doi.org/10.1002/ejsp.302

Kant, I. (1781). *Immanuel Kant's critique of pure reason*. McMillian.

Kelman, H. C. (2018). *Transforming the Israeli-Palestinian conflict: From mutual negation to reconciliation* (P. Matter & N. Caplan, Eds.). Routledge.

Kelman, H. C. (2010). Interactive problem solving: Changing political culture in the pursuit of conflict resolution, peace and conflict. *Journal of Peace Psychology, 16*(4), 389–413. doi:10.1080/10781919.2010.518124

Lazarus, R. (1999). Hope: An emotion and a vital coping resource against despair. *Social Research*, 653–678.

Lee, L. O., James, P., Zevon, E. S., Kim, E. S., Trudel-Fitzgerald, C., Spiro, A., Grodstein, F., & Kubzansky, L. D. (2019). Optimism is associated with exceptional longevity in 2 epidemiologic cohorts of men and women. *Proceedings of the National Academy of Sciences*, 201900712. https://doi.org/10.1073/pnas.1900712116

Leshem, O. A., & Halperin, E. (2020). Hoping for peace during protracted conflict: Citizens' hope is based on inaccurate appraisals of their adversary's hope for peace. *Journal of Conflict Resolution, 64*(7–8), 1390–1417. https://doi.org/10.1177/0022002719896406

Levine, G. F. (1977). "Learned helplessness" and the evening news. *Journal of Communication, 27*(4), 100–105. https://doi.org/10.1111/j.1460-2466.1977.tb01863.x

Litt, M. D., Tennen, H., Affleck, G., & Klock, S. (1992). Coping and cognitive factors in adaptation to in vitro fertilization failure. *Journal of Behavioral Medicine, 15*, 171–187.

Maoz, I., & Shikaki, K. (2014). Joint Israeli Palestinian Poll, December 2014. The Harry S. Truman Research Institute for the Advancement of Peace, The Hebrew University of Jerusalem. https://www.kas.de/c/document_library/get_file?uuid=df9c7294-835c-4ab2-c8ac-a47d15039e28&groupId=252038

Menninger, K. (1959). The academic lecture: Hope. *American Journal of Psychiatry, 116*(12), 481–491.
Mill, J. S. (1875). *Theism* (R. Taylor, Ed.).
Moltmann, J. (1968). Hoping and planning: Future anticipated through hope and planned future. *Cross Currents, 18*(3), 307–318.
Nietzsche, F. (1878/1974). *Human all too human*. Gordon.
Nunn, B. (2005). Getting clear what hope is. In J. Eliott (Ed.), *Interdisciplinary perspectives on hope* (pp. 63–77). Nova Science.
Obama, B. (2006). *The audacity of hope: Thoughts on reclaiming the American Dream*. Random House Large Print.
Olsman, E. (2020). Hope in health care: A synthesis of review studies. In S. C. van den Heuvel (Ed.), *Historical and multidisciplinary perspectives on hope* (pp. 197–214). Springer International. https://doi.org/10.1007/978-3-030-46489-9_11
Peterson, C. (2000). The future of optimism. *American Psychologist, 55*(1), 44–55. https://doi.org/10.1037/0003-066X.55.1.44
Pinker, S. (2011). *The better angels of our nature*. Viking.
Pomper, G. M. (1990). Pessimism, psychology, and politics. *Psychological Inquiry, 1*(1), 68. https://doi.org/10.1207/s15327965pli0101_17
Pruitt, D. G. (1997). Ripeness theory and the Oslo Talks. *International Negotiation, 2*(2), 237–250. doi:https://doi.org/10.1163/15718069720847960
Russell, B. (2001). Developing Derrida pointers to faith, hope and prayer. *Theology, 104*(822), 403–411. https://doi.org/10.1177/0040571X0110400602
Scheier, M., & Carver, C. (1985). Optimism, coping and health: Assessment and implications of generalized outcome expectancies. *Health Psychology, 4*, 219–247.
Schmale, A. H., & Iker, H. (1971). Hopelessness as a predictor of cervical cancer. *Social Science & Medicine (1967), 5*(2), 95–100.
Seligman, M. (1990). *Learned optimism*. Knopf.
Simkin, D. K., Lederer, J. P., & Seligman, M. E. P. (1983). Learned helplessness in groups. *Behaviour Research and Therapy, 21*(6), 613–622. https://doi.org/10.1016/0005-7967(83)90079-7
Smith, A. (2007). Not waiving but drowning: The anatomy of death row syndrome and volunteering for execution. *Boston University Public Interest Law Journal, 17*(2), 237–254.
Snyder, C., Cheavens, J., & Michael, S. T. (1999). Hoping. In C. R. Snyder (Ed.), *Coping: The Psychology of What Works* (pp. 205–227). Oxford University Press.
Snyder, C. R. (1994). *The psychology of hope*. Free Press.
Snyder, C. R. (Ed.). (2000). *Handbook of hope: Theory, measures, and applications* (1st ed.). Academic Press.
Snyder, C. R., Harris, C., Anderson, J. R., Holleran, S. A., Irving, L. M., Sigmon, S. T., Yoshinobu, L., Gibb, J., Langelle, C., & Harney, P. (1991). The will and the ways: Development and validation of an individual-differences measure of hope. *Journal of Personality and Social Psychology, 60*(4), 570.
Snyder, C. R., Rand, K. L., & Sigmon, D. (2005). Hope theory. In C. R. Snyder & S. Lopez (Eds.), *Handbook of positive psychology* (pp. 257–276). Oxford University Press.
Spinoza, B. (1677/2000). *Ethics* (G. H. Parkingson, Trans.). Oxford University Press.
Telhami, S., & Kull, S. (2013). *Israeli and Palestinian public opinion on negotiating a final status peace agreement*. Saban Center at the Brookings Institution.
Thórisdóttir, H., & Jost, J. T. (2011). Motivated closed-mindedness mediates the effect of threat on political conservatism. *Political Psychology, 32*(5), 785–811.
Tiger, L. (1989). *Optimism, the biology of hope*. Simon & Schuster.
Tiger, L. (1999). Hope springs internal. *Social Research, 66*(2), 611–623.

Tillich, P. (1965). *The right to hope*. Harvard Divinity School.

Tops, M., Quirin, M., Boksem, M. A. S., & Koole, S. L. (2017). Large-scale neural networks and the lateralization of motivation and emotion. *International Journal of Psychophysiology, 119*, 41–49. https://doi.org/10.1016/j.ijpsycho.2017.02.004

Zeidner, M., & Allen, L. (1992). Coping with missile attack: Resources, strategies, and outcomes. *Journal of Personality, 60*(4), 709–746. https://doi.org/10.1111/1467-6494.ep9212283734

3 Conceptualizing and Measuring Hope

> Hope is the thing with feathers, that perches in the soul.
>
> **—Emily Dickinson**

Reverend Jesse Jackson made two attempts to be elected as the Democratic Party's nominee for the US Presidency, the first in 1984 and the second in 1988. Both were unsuccessful. Addressing the audience at the 1988 Democratic National Convention in Atlanta, Jackson ended his famous speech with the following words:

> Wherever you are tonight, you can make it. Hold your head high; stick your chest out. You can make it. It gets dark sometimes, but the morning comes. . . . Keep hope alive. Keep hope alive! Keep hope alive! On tomorrow night and beyond, keep hope alive!

Although he knew he had lost when he gave the speech, hope was the main message in Reverend Jackson's powerful address. But what is hope according to Jackson? And what is hope more generally? At first glance, this question might seem odd because hope is an everyday idea used offhandedly with an implicit assumption that hope needs no explanation. Consider the following mundane examples: "I hope this email finds you well," "My hopes for recovery are high because the surgery was successful," "I hope to finish the job on time." People often take it for granted that others know what they mean when they talk about hope. However, looking more closely at the above examples, one discovers that hope is used differently in each sentence.

First, examine the commonplace salutation: "I hope this email finds you well." Here the sender is expressing her *wishes* for the well-being of the receiver. Regardless of whether the sender is genuinely interested in the receiver's well-being or simply writing out of habit, the message conveys the sender's wishes that the receiver is well. Note that there is no hint as to whether the sender *estimates* or *expects* that the receiver is well. The sender's estimation of

Hope Amidst Conflict. Oded Adomi Leshem, Oxford University Press. © Oxford University Press 2024.
DOI: 10.1093/oso/9780197685303.003.0003

the likelihood that the receiver is well is not a part of the sentence's meaning. When people say, "I hope this email finds you well," they use the word hope to signify their wishes or desires rather than their expectations (estimation of likelihood). In sum, "I hope" can mean "I wish."

However, "wish" is not the central meaning conveyed in the following example. When a hospital patient says: "My hopes for recovery are high because the surgery was successful," he means that his expectation (i.e., assessment, estimation) of recovery is higher because the medical operation went well. Here the speaker uses hope to signify an expectation, assessment, or estimation, not necessarily a wish. Of course, the patient clearly wishes to be healthy, but this is not the subject of the sentence because the wish for health existed and continues to exist regardless of the surgery's favorable outcomes. In short, "I hope" can mean "I expect."

What about "I hope to finish the job on time"? Here "hope" can be understood as signifying the speaker's *wishes* to finish the job by the deadline (i.e., "I wish to finish the job on time") or as signifying the speaker's *expectation* that the job will be done by the deadline (i.e., "I expect that I will finish the job on time"). Looking more closely, hope is used here to convey both wishes *and* expectations (i.e., "I wish to finish the job on time, and I expect I can do it"). Indeed, in most cases, hope conveys some combination of a wish (i.e., desire, aspiration) to attain a goal and expectations (assessment of likelihood) that the goal can be attained (Lazarus, 2013; van Zomeren et al., 2019). Of course, wishing to achieve a goal and expecting to achieve it are two different things. One problematic yet fascinating property of hope is that it mixes wishes and expectations in an unknown ratio. How much of one's "hope to finish the job" is about one's desires, and how much of it is one's expectations? It is hard to say.

If the reader is confused, it might be comforting to know that she or he is not alone. Many scholars have failed to see that hope has several meanings. The unfortunate result is that when researchers study hope, they often examine different things. Regrettably, the confusion about hope was, and still is, prevalent in the study of hope during conflict. Consequently, conflict scholars report different, sometimes contradictory findings about hope for peace. I, too, failed to fully grasp the complexities of hope in my first attempts to conceptualize and measure it. This chapter aims to correct the wrongs and rectify the confusion about hope. By the end of the chapter, the reader will have a clearer understanding of how I think hope should be conceptualized and measured. I hope.

In order to conceptualize hope, I examine its *properties.* Focusing on the properties of hope (its structure and anatomy) helps distinguish between elements inherent to the concept and those merely adjunct to it. I use insights from philosophical thought and psychological inquiry to illuminate the properties of hope. Combining the wisdom of the two disciplines is also intended to correct the long-standing problem of disciplinary isolation. Sticking to one discipline means looking for the answers only under the closest lamppost. The answers, however, could be hidden under another streetlight, just a few steps away. Though not an easy task, creating this interdisciplinary dialogue is essential for a more holistic and robust understanding of hope.

The chapter continues as follows. To begin the process of conceptualization, I first outline two fundamental assumptions on hope and then turn to examine its properties with the task of identifying hope's most essential components. Next, I elaborate on common conceptualizations of hope found in the literature and introduce the *bidimensional model of hope*, a novel, parsimonious yet comprehensive conceptualization of hope. Solving discrepancies and inconsistencies in existing literature, the model proposes that wishes and expectations are two perpendicular dimensions of a bidimensional plane. Hope is a place on this plane corresponding with a certain level of wishes to attain a goal and a certain level of expectations of attaining it. Flowing from the conceptualization, I describe how to operationalize hope and exemplify this operationalization on original data collected in Israel-Palestine.

The explorations of the concept of hope presented in the chapter are not relevant only to conflicts. They, and the definitions, models, and measurements of hope I introduce in the following pages, apply to the study of hope in general. Topics like intergroup inequality, social mobility, ontological security, and other contemporary social challenges would benefit from the novel conceptualization and operationalization of hope this chapter offers. Still, the book is about hope amidst conflict, so examples of hope (and its absence) are made in reference to peace and conflict resolution. Focusing on hope in conflict, particularly in intractable conflict, is deliberate. As a scholar of hope, I find it highly useful to examine people's hope in the gloomy conditions of violence and grief. As a conflict scholar, I consider it essential to understand conflict and peace through the prism of hope.

Conceptualizing Hope

Properties of Hope

I begin the examination of hope's properties by mentioning two intuitive assumptions concerning hope. The first thing to say about hope is that it relates to the future. So, when it comes to hope, the future is front and center, the focus of attention (Baumgartner et al., 2008; Fromm, 1968). As a future-oriented construct, hope involves anticipation, imagination, prediction, evaluation, and forecast, but also suspension, uncertainty, and doubt. It is thus not surprising that philosophers and psychologists examine hope alongside other future-oriented constructs like fear (e.g., Halevy, 2017; Jarymowicz & Bar-Tal, 2006; Roseman, 1991; Spinoza, 1677/2000) and despair (e.g., Descartes, 1649/2015; Halperin, 2015; Lazarus, 1999).

The second assumption about hope is that hope concerns something good, favorable, or desirable, at least in the eyes of the hoping person. More accurately, the *target* of hope (i.e., the hoped-for goal or object) is perceived by the hoping person to be positive and good (Averill et al., 1990). One does not hope for something one believes to be undesirable and unwanted. In this regard, hope is the opposite of fear and threat, which concern the anticipation of something hostile and malign (Roseman, 1991).

A clear distinction must be made between the target of hope, which, by definition, is positive in the eyes of the perceiver, and hope itself, which can be harmful (Freud, 1927; Nietzsche, 1878; Spinoza, 1677/2000). As demonstrated in the previous chapter, hope's merits cannot be taken for granted. For Reverend Jackson, hope is a virtue that should be preserved at all costs. For Nietzsche, on the other hand, hope is a curse, a calamity that should be avoided. We can conclude that hope involves the *anticipation that a positive, desired goal will materialize* but that the benefits of this anticipation are not self-evident.

Turning to examine hope's core properties, it is important to note that until the mid-20th century, hope was primarily debated in the arena of theological thought. Often labeled a spiritual matter, hope was deemed unworthy of empirical scrutiny among social scientists (Eliott, 2005; Frank, 1968; Orne, 1968). Scholarly attention was therefore focused on the value of hope rather than its anatomy. The psychiatrist Karl Menninger was one of the first to advocate that hope was an object worthy of empirical research, a "thing" that should be measured and empirically investigated (Menninger, 1959). From the 1960s on, hope was gradually accepted as a valuable topic for scientific examination.

R. S. Downy was one of the first philosophers to examine hope's core properties. According to Downy, the first property intrinsic to hope is a desire to achieve an objective (Downie, 1963). In other words, X hopes for Y only if X wishes for Y (see also Day, 1969). If a person hopes for peace, it can be deduced that he or she wishes for peace. The connection between hope and wishes could also be exemplified in the negative. For example, when one parks in a no-parking zone, one does not *hope* to get a parking ticket because one rarely desires to receive a parking fine. Levels of hope for receiving the parking fine remain null even when the assessment of the likelihood of getting the fine is high (as when seeing a police officer patrolling the street). Therefore, having a desire or a wish to attain a goal is the first prerequisite of hope.

Second, hope implies a perception of possibility, even the slightest one, that the wished-for goal can be attained (Bury et al., 2020; Downie, 1963; Wheatley, 1958). Hope does not exist for things we believe are entirely unattainable. We might *wish* to visit outer galaxies, hug our great-great-grandchildren's children, or become the first political psychologist who is also a NASA astronaut, but hoping so will be a misuse of the concept because the chances to achieve these goals are assumed to be naught. In other words, X has hope for Y only if X believes, even to a minimal extent, in the possibility of Y (Downie, 1963). If Y is considered utterly impossible (like me becoming an astronaut), there is no hope for Y.[1] Having some expectation that a goal can be attained is thus the second property of hope. It is also important to mention that when Y is certain, X does not hope for Y (e.g., one does not hope to get the job after one was accepted to the job). Hope appears to exist between the two extremes of possibility when the chance to reach a goal is not considered completely impossible or inevitable.

Hope, therefore, consists of a wish to attain a goal and an expectation that the goal can be attained. Several social scientists adopted the idea that hope comprises these two components. For example, early psychological research on hope led by Ezra Stotland (1969), examined two factors Stotland believed were intrinsic to hope. The first factor was the "organism's perceived importance of the goal" and the second was the organism's "perceived probability of obtaining that goal." Hope thus depends on how much the goal is perceived as essential or desirable (akin to the *wish* component) and on assessing the probability that this goal can materialize (akin to the *expectation* component). In line with Stotland's research, Averill, Catlin, and Chon showed that the factor distinguishing hope from wish was the expectation of attainment (1990).

Examining hope as a construct that includes two components, wishes and expectations, is also found in the work of Erickson and colleagues (1975), Staats (1989), Lopez et al. (2003), Sagy and Adwan (2006), and Miceli and Castelfranchi (2010). There is also a consensus among hope philosophers that the "standard account" of hope is that "hoping that P" means the "desires for P and the belief that P is possible (but not certain)" (Milona, 2020, p. 101). Admittingly, conceptualizing hope as a construct combining wishes and

[1] I confess that I've always wished to be an astronaut, but I'm quite certain that at this stage NASA will not accept me into their ranks, especially given my fear of flights.

expectations may seem too simple. Yet surprisingly, existing research did not yet propose a coherent model combining the two components.

What are the relationships between wishes and expectations? First, it appears that the two are independent. One can wish to attain a goal with varying degrees of expectations that the goal will be attained. One can also expect to attain a goal with varying degrees of desire. Nevertheless, although the dimensions are independent, there might be some spillover between the two. Perhaps unintentionally, the desire for Y is not entirely isolated from the assessment of the possibility of attaining Y.

For example, people might avoid wishing for something they think is entirely unachievable. In this case, the low expectation influences the wish component of hope. In other instances, people might be motivated to bolster their expectations just because they want something badly. In these "wishful thinking" circumstances, the high wishes affect the expectation component. Indeed, recent studies show that the correlation coefficient between wishes and expectations is significant but that the value of this correlation is low to moderate (Pearson's *r* between .1 and.3) (Leshem, 2017, 2019; Leshem & Halperin, 2020b). This means that wishes and expectations commonly go together but that the link between them is loose. It looks like each dimension enjoys independence and, as such, should be explored separately. However, it also appears that there might be some influence of each dimension over the other, so the relations between the two components should also receive attention.

I have now argued that wishes and expectations are two distinct (though correlated) dimensions of hope. Nevertheless, the colloquial use of the word hope blurs this distinction, like in the example presented earlier "I hope to finish the job on time." To further demonstrate this confusion, think about the following sentence I once heard from a close Israeli friend: "I really hope there will be peace, but my hope for peace is really low." Most readers easily understand that my friend expressed her high wishes for peace and low expectations that peace would materialize. The context helped you, the reader, change the interpretation of hope from wishes to expectations and understand this seemingly contradicting statement in which my friend simultaneously expressed high and low hopes for peace.[2]

The main arguments proposed so far can be summed in the following way. First, hope comprises at least two distinct but interrelated components: a wish to attain a goal and an expectation (though not certainty) of attainment.

[2] The confusion is not confined only to English. In other languages, "hope" can also mean "wish," "expect," or some combination of both.

Second, when people use the word "hope," they use it to signify the wish component, the expectation component, or an amalgam of both. Are there other components essential for hope? Are there alternative conceptualizations?

Hope as an Emotion

Many psychologists study hope as an emotion (e.g., Cohen-Chen et al., 2015; Halperin, 2015; van Zomeren, 2021). Richard Lazarus, for example, one of the founding fathers of appraisal theory, maintains that hope functions, first and foremost, as an emotion (1999). Appraisal theory posits that emotions are products of rapid, often unconscious, appraisals of situations. After the situation is quickly appraised, a discrete emotion that corresponds with the appraisal is elicited. The emotion then becomes a primary factor governing further cognitive, emotional, and behavioral processes. The emotion of hope is evoked based on two appraisals: the first is the judgment of whether the situation is desired, and the second concerns the certainty of the situation (Smith & Ellsworth, 1985). The emotion of hope will be elicited when a future situation is judged as desirable and possible (but not certain).

Conceptualizing hope as an emotion elicited by quick appraisals is used by most contemporary scholars who investigate hope in the context of conflicts (Cohen-Chen, Halperin, Crisp, et al., 2014; Cohen-Chen et al., 2015; Halperin, 2015; Hasan-Aslih et al., 2019; Rosler et al., 2017). There are, however, significant challenges when examining hope as an emotion. The first is that, like other emotions, the emotional experience of hope is transient and short-termed (Halperin, 2016; Itkes & Kron, 2019; Robinson & Clore, 2002a, 2002b). People might experience the emotion of hope when they learn about a new opportunity or receive a notice that they were shortlisted for a job interview. In the context of conflict, people might experience hopeful emotions when they read about a breakthrough in negotiations or see the handshakes of leaders who used to be enemies. In these moments, the situation is appraised as highly desirable and the chances of success as possible, which together provoke the emotional response of hope. However, after a while, the emotional experience dissipates and eventually disappears. The fading of the emotional experience does not mean that the appraisals have changed or that we have lost "hope." It simply means that the emotional experience is negligent or gone.

According to Robinson and Clore (2002a, 2002b), the transient nature of emotions requires that we measure them when or immediately after they are experienced. This way, the self-reported measure captures the emotion as closely as possible. Nevertheless, after the emotional experience dissipates, answers will likely reflect nonaffective constructs. Put differently, when we try to measure emotions after they fade, we are likely to capture something other than emotions. Itkes et al. (2017) and Itkes and Kron (2019) demonstrate that one such nonaffective construct is semantic knowledge of *valence*, which is the positive or negative sign we ascribe to an object.

The above implies that if we want to study the emotional aspect of hope during conflict (which is undoubtedly a worthwhile endeavor), we need to measure hope when it is activated. This can be done by deliberate activation (in an experimental study) or tapping into people's emotional hope experiences in times of great exhilaration. However intriguing, these approaches are difficult to implement and provide only a partial picture of hope for peace. As scholars of conflict, we are interested to know the extent people hope for peace across populations, eras, and contexts, even when the emotional experience is dormant. The study of hope as an emotion does not allow this.

Looking at hope as a sentiment rather than an emotion addresses the problem of the short duration of emotions.[3] Sentiments are a "disposition to respond emotionally to a certain object" (Frijda et al., 1991, p. 207). In other words, sentiments are long-lasting and lingering emotional states. In research on protracted conflicts, anger toward the adversarial group has been studied as a sentiment due to anger's stable presence and long-standing effects (Halperin & Gross, 2011). Intergroup hate has been conceptualized as an emotion and a sentiment because it can occur on the immediate and chronic levels (Fischer et al., 2018). Building on this logic, hope may also be defined as an emotion and a sentiment because it is sometimes experienced immediately but can evolve into a lingering disposition if experienced repeatedly and over time (Cohen-Chen et al., 2017; Halperin et al., 2011).

However, defining hope as a sentiment does not solve the problem of its empirical exploration because the study of hope as an emotion or sentiment is based on self-reported measures (e.g., Cohen-Chen, Halperin, Crisp et al., 2014; Hasan-Aslih et al., 2019; Rosler et al., 2017). For example, seeking to gauge the emotional or sentimental reaction, research on hope in conflict often includes questions asking respondents to rate the extent to which they "feel hope regarding future peace." However, the response to this question is likely to capture the extent participants believe peace is possible, not necessarily the emotional or sentimental reactions to this appraisal (Itkes & Kron, 2019). Other self-reported questions about hope for peace might gauge the extent to which people wish for peace, not the emotion or sentiment associated

[3] I wish to thank Eran Halperin for highlighting this point.

with that wish. More generally, research on hope as an emotion might unintentionally measure the appraisals that elicited the emotion, not the emotion itself.

To further complicate the problem, hope can signify wishes, expectations, or their combination. Thus, when asked about how much hope they feel when thinking about peace, respondents might answer about the intensity of their wishes for peace, their expectations for peace, or some unknown combination of the two dimensions. In fact, using the word "hope" in self-reported measures is a problem regardless of whether hope is defined as an emotion, sentiment, cognition, or any other way. When asked, "How much hope do you have for peace?" how can researchers know whether participants report their wishes for peace or their expectations that peace will materialize (or some mixture of the two)? The inherent fuzziness of the word "hope" creates a real empirical headache.

Consider the following studies. In their research on hopes for peace in Israel-Palestine, Rosler, Cohen-Chen, and Halperin (2017) reported low levels of hope for peace among Jewish Israelis (2.9 on a scale from 1 to 6). Low levels of hope were also reported in 1976 by 59.1% of Israeli Jews who expressed hopelessness about the chances of peace (Stone, 1982). However, several studies show that Jewish Israelis have *high* hopes for peace. A study published in 2008 found that Jewish Israelis' hope for peace was 4.9 on a scale of 1 to 6 (Halperin et al., 2008).[4] Many years earlier, Antonovsky and Arian (1972) also reported high hopes for peace among Israeli Jews. According to the study, "hope for peace with the Arabs" scored higher than hope for national prosperity and economic stability put together.

The research mentioned above reveals mixed results. Some studies report low hopes for peace among Jewish Israelis (e.g., Rosler et al., 2017; Stone, 1982; Telhami & Kull, 2013), while others report their high hopes for peace (e.g., Antonovsky & Arian, 1972; Halperin et al., 2008). These differences may be attributed to the different political circumstances present when the data were collected. However, scrutinizing the studies, it seems that the researchers were measuring different things. For example, Stone (1982) and Rosler et al. (2017), whom both reported on Jewish Israelis' low "hopes" for peace, may have measured participants' *expectations* for resolution, which might explain the low levels of "hope". Similar issues can be found in studies on hope where the authors report on participants' hopes for peace but may have gauged only the expectation dimension (Cohen-Chen et al., 2015,

[4] I transformed the original scale to improve comparison.

2016; Cohen-Chen, Halperin, Crisp, et al., 2014; Cohen-Chen, Halperin, Porat, et al., 2014).

On the other hand, Antonovsky and Arian (1972) also asked participants to report their "hopes for peace" but were de facto gauging participants' *wishes for peace*, not their expectation whether peace would materialize. Unsurprisingly, here the levels of hope for peace were high. Similar issues are present in a study conducted on the hopes of Palestinian citizens of Israel to advance intergroup harmony versus intergroup equality between the Jewish Israeli majority and the Palestinian minority in Israel (Hasan-Aslih et al., 2019). The researchers compared participants' "hopes" for the two advancements, but may have compared only participants' wishes for them. Last, in Halperin et al.'s study (2008), it is unclear if respondents expressed their wishes for peace, their assessment of its feasibility, or both. Overall, it seems that the different meanings of hope in everyday speech affected the research on hope during conflict. All researchers report on "hope" but, some report on wishes, others on expectations, while others report on an unknown blend of the two dimensions.

The most important conclusion drawn from these examples is that it is too risky to use the word "hope" when it comes to measuring it. If hope signifies two different things, how can scholars determine which one they are measuring? Researchers cannot even be confident that they are measuring their "mixture." By avoiding the confusing word "hope" and instead focusing on its discrete dimensions, scholars of hope can curb the problem and, in fact, provide a much more exciting and meaningful account of hope. Before I describe a new way to conceptualize hope, I briefly discuss the connection between hope and action.

Hope as Action

Rick Snyder proposed that hope entails more than wishes and expectations because it must contain commitment and a plan of action (Snyder, 1994). In other words, if X hopes for Y, X must try to achieve Y. According to Snyder and other hope scholars (Averill et al., 1990; Breznitz, 1986), wishes and expectations are necessary but not the only core components of hope. Some act or intention to act to attain the goal is required. According to this approach, hoping for something without intending to do anything about it cannot be called hoping. The intentional or behavioral component is thus part and parcel of hope, not merely its product (Bloch, 1959; Fromm, 1968; Snyder et al., 2005; Tillich, 1965).

Some empirical evidence points to the vital link between hope and action. Averill and colleagues examined students' self-reported reactions to objectives they hoped for compared to objectives they *wished* for (Averill et al., 1990). When describing their reactions to things they *hoped* for, participants reported that they "work hard" and "become better organized." However, descriptions of activity were not found when participants reported their reactions to things they wished for. The authors conclude that hope is conceptually linked to activity. Similarly, psychologist Shlomo Breznitz, who studied hope among individuals under stressful conditions, adds that "In order for hope to 'work,' it must not be idle, but active" (1986, p. 295). Other empirical work connecting hope and collective action for social change highlight the strong link between hope and action (Greenaway et al., 2016; Leshem, 2019).

Looking at intention, commitment, and action as core components of hope resonates with the definitions of hope of thinkers like Ernest Bloch, Victor Frankl, Eric Fromm, and Paul Tillich. These thinkers maintained that hope could only be manifested in action. This approach argues that the concept of hope is misused if wishes and expectations are not accompanied by deliberation, intention, commitment, or action. As stated by Fromm, if one wishes for salvation or liberation, a satisfying life, or peace for all humanity but does nothing but wait for it to come, one is not engaged in true hoping (Fromm, 1968). Even if hope resides in an individual's innermost thoughts, it can only be judged in the form of action, for hope is not present without an activity aimed at achieving the goal (Bloch, 1959).

Psychologists like Snyder and others propose that hope's behavioral aspect is an innate part of hope, not merely its consequence. However, it is worth noting that this approach diverges from the common psychological distinctions between mental states and behaviors. Most of the time, emotions, beliefs, and other mental constructs are separated from their behavioral results. Social psychologists often study mental states and processes and explore when and how they impact behavior (sometimes also investigating the reverse causality). Although hope without intention to act can hardly be effective, I stick to the common notion that mental states are distinct from their behavioral outcomes. Therefore, in this book, I separate hope from the intention to act upon this hope. I now turn to offer a novel way to think about hope and measure it.

The Bidimensional Model of Hope

So far, I have broadened the exploration of hope by examining its properties and the connection between them. To increase the scope and depth of this exploration, I used the prisms of philosophy and psychology and considered several approaches to conceptualize hope. Maybe the most consistent idea conveyed through this exploration is that hope involves the combination of wishes and expectations. Quite straightforwardly, one hopes when one wishes to attain a goal and expects it to materialize. Note that "hope as an emotion" and "hope as a behavior" discussed earlier can be viewed as consequences of wishing and expecting, not as core components of hope. Appraisal theory suggests that hopeful emotions stem from wishing for something and having some expectations of attainment (e.g., Cohen-Chen, Halperin, Porat, et al., 2014; Erickson et al., 1975; Lazarus, 1999). Intention, commitment, and actions aimed at attaining the goal also derive from the primary elements of wishes and expectations. I suggest that if hope scholars seek to focus on the essence of hope and not take the risk that they are studying its derivatives, then they must focus on wishes and expectations.

Another reason we can feel confident in defining hope as the interplay between wishes and expectations is that all conceptualizations of hope are based, in some form, on the presence of these two components. Across the many approaches to conceptualizing hope, none contradict the idea that hope requires both the wish to attain a goal and the expectations of attaining it. Even emotional approaches to hope build on the notion that emotional experiences of hope are based on quick appraisals of wishes and expectations. Philosophers and popular dictionaries also define hope as the combination of the two components (see Eliott, 2005). Webster's *Dictionary*, for example, defines the noun "hope" as a "*desire* accompanied by *expectation* of or belief in fulfillment," Oxford's *Advanced Learner's Dictionary* defines the verb "hope" as "to *want* something to happen and *think that it is possible*."

After establishing that hope comprises wishes and expectations, the next natural question is how these two components compile into "hope." Do wishes and expectations add up to create hope? Is hope a product of their multiplication? Calculating hope by adding or averaging wishes and expectations does not make sense because when one of the components equals zero, hope should remain zero even when the level of the other component is high. For example, if there is no desire but some expectation, aggregating or averaging the two would yield a number greater than zero, which does not make sense because hope does not exist without desire (recall the parking ticket example). Multiplying wishes and expectation solves this problem and is in line with work on motivation that uses the product of the perceived value of a goal and the perceived expectancy to achieve a goal as a predictor of behaviors pursuing the goal (Forster et al., 2007; Kruglanski et al., 2015). Multiplying wishes and expectations is also in line with economic models of expected utility (Schoemaker, 1982), where the prospect's value is multiplied by the prospect's chances of materializing. Yet there are two fundamental problems with conceptualizing hope as the product of wishes and expectations.

The first is that multiplying assumes equal weight to wishes and expectations. However, the weight of each dimension—namely, its importance or relevance for hope—is context-dependent. In some contexts, hope might depend more on one's wishes to achieve a goal than one's expectations of achieving it, and, in other cases, one's expectations may be more influential than one's wishes. The second related problem is that wishes and expectations are not interchangeable. For instance, the hopes of a person with low wishes and high expectations to achieve a goal are qualitatively different from those of someone with high wishes and low expectations for the same goal. The first person is not very enthusiastic about the goal but thinks it is possible, while the second is very eager about the goal but thinks attainment is unlikely. Their hopes for their goal are qualitatively different even though the product of the components is identical.

The bidimensional approach I use to conceptualize hope (see Figure 3.1) bypasses this predicament by mapping hope on two orthogonal dimensions, with one dimension being the wish to attain a goal and the other being the expectations of attaining it (for a similar approach for conceptualizing socio-psychological constructs as bidimensional see Fiske [2018] and Fiske et al. [2007]). Thus, instead of summing or multiplying the two components to create a single value, the bidimensional model creates a much more nuanced picture with hope corresponding with two values: wishing and expecting. This approach is exhaustive enough to encompass the many manifestations of hope and parsimonious enough to include only those elements that are completely essential.

I now examine the possible intersections between wishes and expectations, as presented in Figure 3.1. Hope is practically nonexistent when wishes and expectations are close to zero (bottom-left corner, outside the shaded area). One simply does not have hope for things one does not wish for or thinks will never happen.[5] Moving upward and to the right, we leave the extremely low levels of wishes and expectations and enter the shaded area representing hope in all its diverse combinations and magnitudes. As wishes and expectations

[5] Note that not wishing for X does not necessarily mean wishing for the opposite. For example, when someone admits that he has no wishes for peace, that does not necessarily mean that one wishes for war. Similarly, not expecting X to materialize does not necessarily mean expecting the opposite to materialize.

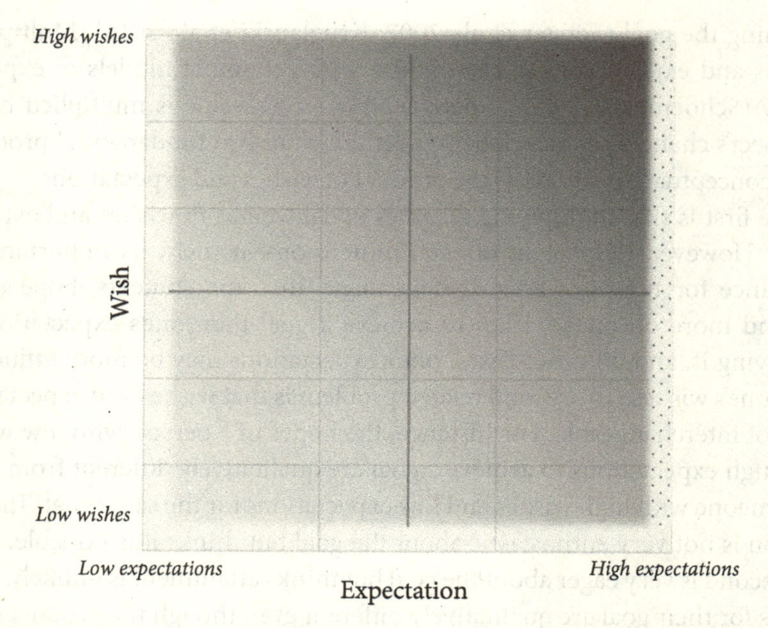

Figure 3.1 The Bidimensional Hope Plane. Shaded area represents the presence of hope (the darker the color, the higher the hope). The doted area represents an absence of hope due to certainty.

rise (i.e., the closer we move to the upper-right corner), the higher the hope. Hope is at its peak when wishes and expectations are both very high, as represented in the upper-right corner. This happens when things one greatly desires seem increasingly feasible. However, when expectations reach the levels of certainty (dotted area), hope is no longer relevant. During certainty, we exit the area of hope because one does not hope for things one knows will surely transpire.

Looking at the shaded area, it is pretty clear that levels of hope are low in the lower-left section and high in the upper-right section. But what happens in the other two sections of the shaded area? I first examine the lower-right section, where the wishes to attain a goal are somewhat low, but the expectations are high. This section represents instances when people are not too enthusiastic about a goal but believe this goal can be attained. Optimism might describe this corner quite well because, compared to hope, optimism commonly involves targets that are less important but perceived to be more attainable (Bruininks & Malle, 2005; Halperin, 2015). We can express optimism about things we are not enthusiastic about but believe their chances of fulfillment are high. Similarly, Bury et al. (2018) reveal that hope and optimism differ in the expectations dimension. When the likelihood of the desired goal is high, optimism and hope are used interchangeably. However, when the likelihood of the desired goal is low, "hope" rather than "optimism" was the preferred term to describe the situation.[6]

Now examine the upper-left section, where wishes are high but expectations low. This corner is probably the most relevant area of hope for our inquiry because, during conflict, peace is desired (at least for most people most of the time) but believed to be unobtainable. For this reason, most of the chapters of this book are devoted to the high-wish low-expectations corner. Bury et al. (2020) and other thinkers and public figures also focused on this type of hope, where we desire something badly but believe that the chances of achievement are slim.

I wish to address several, perhaps related, points that may have occurred to the reader by now. The first point concerns the opposite of hope. In some literature, fear is mentioned as antithetical to hope because the two involve envisioning a different future, where hope is about a positive future and fear a negative one (Jarymowicz & Bar-Tal, 2006; Roseman, 1991). Other research mentions despair as the opposite of hope (Halperin, 2016; Nesse, 1999). This claim also seems plausible because hope involves high prospects of attaining a desired goal, whereas despair involves low prospects of attaining the desired goal. Fear and despair or not synonymous. So, which one is the opposite of hope?

The bidimensional approach quickly solves this puzzle. Quite simply, both are the opposites of hope (see also: Blöser, 2020; Day, 1969). Despair is elicited when a goal, event, or object is appraised as highly desirable but the chances of attainment (i.e., expectations) are appraised in the negative. Simply put, we wish very much but know attainment is impossible. On the bidimensional plane, despair will be located to the left of the Y-axis (demonstrating negative expectations) and above the X-axis (signifying a positive wish). Fear, on the other hand, is evoked when probabilities are appraised as high, but the event or object is appraised as undesirable. We do not want something undesired to happen, but we believe it will. The more undesired the event and the more likely it is, the more we fear it. Placing fear on the bidimensional plane will be to the right of the Y-axis (demonstrating positive expectations) but beneath the X-axis (signifying a negative wish).

The second point concerns the sequence of the dimensions. What comes first, our desires or expectations? Is there some mental order? Is one processed

[6] Another commonly made distinction between hope and optimism is that optimism is "hope minus a plan" (Snyder, 2000). In other words, optimism does not require intention, commitment, planning, or action.

before the other? It could be postulated that appraising the desirability of a goal precedes the appraisals about its feasibility, though no direct evidence for this claim was provided. A related point concerns the hierarchy between the dimensions. Is one dimension more important than the other, and should it thus receive more attention or weight? If so, is this hierarchy maintained across all contexts? And what exactly are the criteria for determining which dimension is more important? Questions of sequence and hierarchy are essential to our inquiry but will be answered in the following chapters.

Operationalizing Hope

Studies that explored hope during conflict used different approaches to operationalize hope. I have compiled these approaches in a table in the Appendix. The table presents twenty-six empirical sources (journal articles and book chapters) that aimed to measure hope during conflict. The table includes the definition of hope used in each source, the items used to operationalize hope, and a short description of the sample. Most importantly, the table includes an attempt to estimate which dimensions of hope were measured.

As seen in the Appendix, though all researchers report on "hope," the conceptualization and operationalization differ significantly from one study to the next. Moreover, due to the fuzziness of hope, it is hard to know how participants interpreted the word "hope" in many of these studies. Did respondents think they were asked about their "wishes," "expectations," or some combination of both? In some publications, it is possible to provide a good guess as to how "hope" was interpreted by participants. In other cases, it is quite impossible. On top of that, it is plausible that the researchers focused and reported on one dimension (still referring to it as "hope") but that participants interpreted the concept based on the other dimension. Another problem is that, in some studies, how hope was operationalized does not derive from how it was conceptualized. A look at the twenty-six studies compiled in the Appendix quickly leads to the conclusion that research on hope would greatly benefit from a more standardized approach to operationalizing hope.

From Concept to Measurement

Since 2017, I have operationalized hope based on the bidimensional model described above (Leshem, 2017, 2019; Leshem & Halperin, 2020a, 2020b,

2021). Quite straightforwardly, to estimate the levels of hope for X, we need to assess the degree that people wish for X and the degree that people expect X. For example, using self-reported items, we can ask how much, on a scale from 1 to 6, one *wishes* to reconcile with the enemy and then ask how much one *expects* to reconcile with the enemy. We then place the answers on a bidimensional plane to reveal the hopes for X—in this case, reconciling with the enemy.

So far the bidimensional model proved to be highly fruitful. Not only does it solve inconsistencies stemming from hope's different interpretations, but it also creates exciting avenues for research. For instance, instead of comparing the "hopes" for peace between rival parties or across time, we can compare the wish for peace across the parties or time and then conduct an independent comparison between the expectations for peace of each party or time (Sagy & Adwan, 2006). This approach was used in a recent study conducted in Israel-Palestine (Leshem & Halperin, 2020a). In the study, we found that, all else equal, Palestinians' and Israelis' wish for peace is identical, but Palestinians had higher expectations that peace could be achieved. It seems that both populations equally and strongly desire peace, but, compared to Israelis, Palestinians think future peace is more likely—more about these findings in Chapter 4. We can also investigate the differences between wishes and expectations and detect the factors influencing the congruence between the dimensions. We can then examine whether people with congruent wishes and expectations think differently about conflict and peace than those with discordant wishes and expectations.

Moreover, we can conduct a much more nuanced study of the demographic and sociopolitical antecedences of hope by looking at the potential predictors of each dimension. Studies by my colleagues and I show that certain factors predict how much people wish for peace while others predict how much they expect peace to materialize (see Chapter 4). Most importantly, we can explore the role of each dimension as a predictor of attitude and behavioral change in support of peace and conflict resolution and then take this knowledge to design effective interventions (Chapter 7). The bidimensional approach can inform whether it is more effective to increase the desirability of peace or the belief in its feasibility in promoting conciliatory attitudes among people involved in conflict. These are just a few of the many research possibilities generated when hope is conceptualized and operationalized as a bidimensional construct; some of these avenues are presented later in the book. At this stage, I exemplify the utility of the bidimensional model on data collected among Jewish Israelis and Palestinians.

Applying the Bidimensional Model: Measuring Hope for Peace in Israel-Palestine

How hopeful are Israelis and Palestinians about future peace? In the summer of 2017, I conducted the first round of the Hope Map Project, a global study revealing the antecedences and consequences of hope across conflict zones. The first round was conducted in Israel-Palestine among 500 Jewish Israelis from Israel and 500 Palestinians from the West Bank and Gaza Strip. In the following chapters, I elaborate on the administration of the project, its many empirical and ethical challenges, and counterintuitive results. Here, I want to focus on one item demonstrating the bidimensional model's usefulness.

In this study, all participants were asked to rate how much they wished for a "peace accord that will address the core needs of the two peoples." This phrasing offers a very basic and generic definition of peace, yet it emphasizes its reciprocal nature, such that the needs of both people must be met. Several questions later, participants were asked to assess the chances that a "peace accord that will address the core needs of the two peoples" would indeed materialize. Results are plotted in Figures 3.2 and 3.3.

The figures show similar trends across the rival societies (with slight differences I elaborate on later). First, most Palestinians and Israelis seem to fall in the upper left section, where wishes for peace are high, but expectations for peace are low. This pattern is not surprising. It stands to reason that people

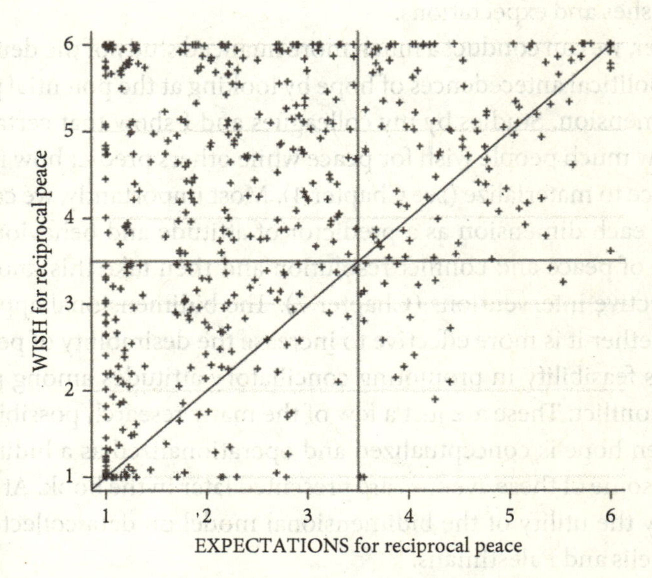

Figure 3.2 Wishes and expectations for peace (reciprocal): Palestine.

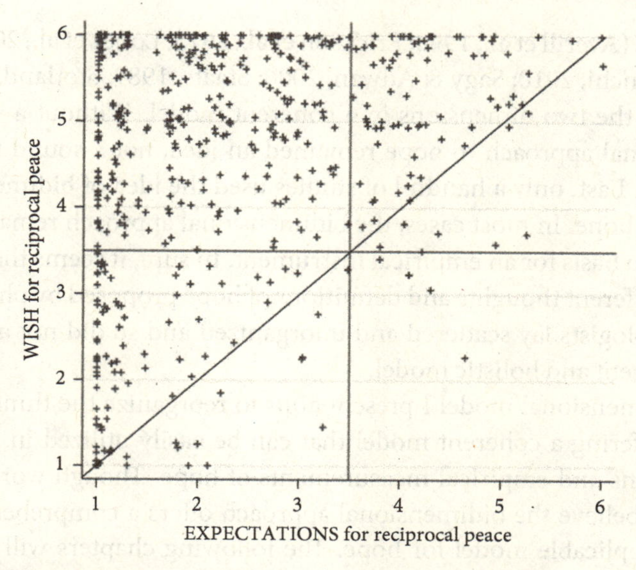

Figure 3.3 Wishes and expectations for peace (reciprocal): Israel.

engulfed in decades of conflict want peace but believe it will never come. Indeed, the average wish for peace across the two societies was moderately high (mean [M] = 4.37 [on a scale from 1 to 6], standard deviation [SD] = 1.45), while the expectation for peace was relatively low (M = 2.41 [on a scale from 1 to 6], SD = 1.32). Furthermore, the hopes of only a few participants fell beneath the diagonal line that represents equal wishes and expectations. It appears that, for most people (85.2% Jewish Israelis and 71.4% Palestinians), levels of desires for peace surpass levels of expectations.

Another important finding is that the correlations between wishes and expectations are significant but moderate ($r = 0.28, p < .001$). This finding aligns with the claim that the dimensions are associated but also highly independent. This is true in the Palestinian sample ($r = 0.32, p < .001$) and the Israeli sample ($r = 0.29, p < .001$).

The above analysis exemplifies some of the utilities of operationalizing hope as a bidimensional construct. Nevertheless, a question arises as to why the bidimensional approach was not offered and utilized throughout the years of hope research. I suggest that at least three factors prevented the bidimensional approach from emerging in social science research. The first is that "hope" is a fuzzy word, and this inherent fuzziness cannot be easily wrung out. In many studies, the colloquial use of "hope" misguided scholarly explorations, and so the distinction between the two dimensions was hidden from the eyes of many. Next, even hope scholars who did distinguish between the two dimensions (Averill et al., 1990; Erickson et al., 1975; Lopez et al., 2003; Miceli & Castelfranchi, 2010; Sagy & Adwan, 2006; Staats, 1989; Stotland, 1969) did not embed the two dimensions in a coherent model. Without a model, the bidimensional approach to hope remained an idea, not a sound theoretical framework. Last, only a handful of studies used the idea of bidimensionality to measure hope. In most cases, the bidimensional approach remained theoretical, not a basis for an empirical instrument. In sum, it seems that, up until now, the different thoughts and definitions of hope proposed by philosophers and psychologists lay scattered and unorganized and so did not accumulate into a coherent and holistic model.

The bidimensional model I present aims to reorganize the thinking about hope by offering a coherent model that can be easily utilized in theoretical examinations and empirical measurements of hope. Though work is still to be done, I believe the bidimensional approach offers a comprehensive, flexible, and applicable model for hope. The following chapters will show how exploring hope through its distinct dimensions generates new insights and novel avenues for research. Chapter 4 illustrates that each dimension has different antecedences. Chapter 5 reveals that peace activists discern between the dimensions in their work. Chapter 6 demonstrates that leaders' speeches contain the two dimensions of hope but to different extents depending on the power structure of the conflict, and Chapter 7 reveals which dimension is more influential when it comes to predicting support for peacebuilding. Perhaps most importantly, the applicability of the bidimensional model of hope is not confined to the context of intergroup conflict. The study of any political and social phenomenon can benefit from this approach. Separately exploring hope's wish and expectation dimensions can inform research on many topics, from the study of public opinion on equality to the study of elite decision-making. The next chapter is devoted to the determinants of the two dimensions of hope for peace.

References

Antonovsky, A., & Arian, A. (1972). *Hope and fears of Israelis*. Jerusalem Academic Press.

Averill, J., Catlin, G., & Chon, K. K. (1990). *Rules of hope*. Springer-Verlag.

Baumgartner, H., Pieters, R., & Bagozzi, R. P. (2008). Future-oriented emotions: Conceptualization and behavioral effects. *European Journal of Social Psychology*, *38*(4), 685–696. https://doi.org/10.1002/ejsp.467

Bloch, E. (1959). *The principle of hope* (vol. 1–3). MIT Press.

Blöser, C. (2020). Enlightenment views of hope. In S. C. van den Heuvel (Ed.), *Historical and multidisciplinary perspectives on hope* (pp. 61–76). Springer International. https://doi.org/10.1007/978-3-030-46489-9_4

Breznitz, S. (1986). The effect of hope on coping with stress. In M. H. Appley (Ed.), *Dynamics of stress* (pp. 295–306). Plenum.

Bruininks, P., & Malle, B. F. (2005). Distinguishing hope from optimism and related affective states. *Motivation and Emotion, 29*(4), 324–352. https://doi.org/10.1007/s11031-006-9010-4

Bury, S. M., Wenzel, M., & Woodyatt, L. (2018). Confusing hope and optimism when prospects are good: A matter of language pragmatics or conceptual equivalence? *Motivation and Emotion*. https://doi.org/10.1007/s11031-018-9746-7

Bury, S. M., Wenzel, M., & Woodyatt, L. (2020). Against the odds: Hope as an antecedent of support for climate change action. *British Journal of Social Psychology, 59*(2), 289–310. https://doi.org/10.1111/bjso.12343

Cohen-Chen, S., Crisp, R. J., & Halperin, E. (2015). Perceptions of a changing world induce hope and promote peace in intractable conflicts. *Personality and Social Psychology Bulletin, 41*(4), 498–512. https://doi.org/10.1177/0146167215573210

Cohen-Chen, S., Crisp, R. J., & Halperin, E. (2016). Hope comes in many forms: Out-group expressions of hope override low support and promote reconciliation in conflicts. *Social Psychological and Personality Science, 8*(2), 153–161, 1948550616667612. https://doi.org/10.1177/1948550616667612

Cohen-Chen, S., Crisp, R. J., & Halperin, E. (2017). A new appraisal-based framework underlying hope in conflict resolution. *Emotion Review, 9*(3), 208–214. https://doi.org/10.1177/1754073916670023

Cohen-Chen, S., Halperin, E., Crisp, R. J., & Gross, J. J. (2014). Hope in the Middle East: Malleability beliefs, hope, and the willingness to compromise for peace. *Social Psychological and Personality Science, 5*(1), 67–75. https://doi.org/10.1177/1948550613484499

Cohen-Chen, S., Halperin, E., Porat, R., & Bar-Tal, D. (2014). The differential effects of hope and fear on information processing in intractable conflict. *Journal of Social and Political Psychology, 2*(1), 11–30. https://doi.org/10.5964/jspp.v2i1.230

Day, J. P. (1969). Hope. *American Philosophical Quarterly, 6*, 89–102.

Descartes, R. (1649/2015). *The passions of the soul and other late philosophical writings*. Oxford University Press.

Downie, R. S. (1963). Hope. *Philosophy and Phenomenological Research, 24*(2), 248–251. https://doi.org/10.2307/2104466

Eliott, J. (2005). What have we done with hope? A brief history. In J. Eliott (Ed.), *Interdisciplinary perspectives on hope* (pp. 3–46). Nova Science.

Erickson, R., Post, R., & Paige, A. (1975). Hope as a psychiatric variable. *Journal of Clinical Psychology, 31*, 324–329.

Fischer, A., Halperin, E., Canetti, D., & Jasini, A. (2018). Why we hate. *Emotion Review, 10*(4), 309–320. https://doi.org/10.1177/1754073917751229

Fiske, S. T. (2018). Stereotype content: Warmth and competence endure. *Current Directions in Psychological Science, 27*(2), 67–73. https://doi.org/10.1177/0963721417738825

Fiske, S. T., Cuddy, A. J. C., & Glick, P. (2007). Universal dimensions of social cognition: Warmth and competence. *Trends in Cognitive Sciences, 11*(2), 77–83. https://doi.org/10.1016/j.tics.2006.11.005

Forster, J., Liberman, N., & Friedman, R. S. (2007). Seven principles of goal activation: A systematic approach to distinguishing goal priming from priming of non-goal constructs. *Personality and Social Psychology Review, 11*(3), 211–233. https://doi.org/10.1177/1088868307303029

Frank, J. (1968). The role of hope in psychotherapy. *International Journal of Psychotherapy*, *5*(5), 383–395.

Freud, S. (1927). *The future of an illusion*. Hogarth.

Frijda, N., Mesquita, B., Sonnemans, J., van Goozen, S., & Strongman, K. (1991). The duration of affective phenomena or emotions, sentiments and passions. In Stongman, K. T. (Ed.), *International review of studies on emotion* (pp. 187–225). Wiley.

Fromm, E. (1968). *The revolution of hope*. Harper & Row.

Greenaway, K. H., Cichocka, A., van Veelen, R., Likki, T., & Branscombe, N. R. (2016). Feeling hopeful inspires support for social change. *Political Psychology*, *37*(1), 89–107. https://doi.org/10.1111/pops.12225

Halevy, N. (2017). Preemptive strikes: Fear, hope, and defensive aggression. *Journal of Personality and Social Psychology*, *112*(2), 224–237. https://doi.org/10.1037/pspi0000077

Halperin, E. (2015). *Emotions in conflict: Inhibitors and facilitators of peace making*. Routledge.

Halperin, E. (2016). *Emotions in conflict: Inhibitors and facilitators of peace making*. Routledge.

Halperin, E., Bar-Tal, D., Nets-Zehngut, R., & Drori, E. (2008). Emotions in conflict: Correlates of fear and hope in the Israeli-Jewish society. *Peace and Conflict: Journal of Peace Psychology*, *14*(3), 233–258. https://doi.org/10.1080/10781910802229157

Halperin, E., & Gross, J. J. (2011). Intergroup anger in intractable conflict: Long-term sentiments predict anger responses during the Gaza War. *Group Processes & Intergroup Relations*, *14*(4), 477–488. https://doi.org/10.1177/1368430210377459

Halperin, E., Sharvit, K., & Gross, J. J. (2011). *Emotion and emotion regulation in intergroup conflict*. In D. Bar-Tal (Ed.), *Intergroup Conflicts and Their Resolution* (pp. 83–103). Psychology Press.

Hasan-Aslih, S., Pliskin, R., van Zomeren, M., Halperin, E., & Saguy, T. (2019). A darker side of hope: Harmony-focused hope decreases collective action intentions among the disadvantaged. *Personality and Social Psychology Bulletin*, *45*(2), 209–223. https://doi.org/10.1177/0146167218783190

Itkes, O., Kimchi, R., Haj-Ali, H., Shapiro, A., & Kron, A. (2017). Dissociating affective and semantic valence. *Journal of Experimental Psychology: General*, *146*(7), 924–942. https://doi.org/10.1037/xge0000291

Itkes, O., & Kron, A. (2019). Affective and semantic representations of valence: A conceptual framework. *Emotion Review*, *11*(4), 283–293. https://doi.org/10.1177/1754073919868759

Jarymowicz, M., & Bar-Tal, D. (2006). The dominance of fear over hope in the life of individuals and collectives. *European Journal of Social Psychology*, *36*(3), 367–392. https://doi.org/10.1002/ejsp.302

Kruglanski, A. W., Jasko, K., Chernikova, M., Milyavsky, M., Babush, M., Baldner, C., & Pierro, A. (2015). The rocky road from attitudes to behaviors: Charting the goal systemic course of actions. *Psychological Review*, *122*(4), 598–620. https://doi.org/10.1037/a0039541

Lazarus, R. (1999). Hope: An emotion and a vital coping resource against despair. *Social Research*, *66*(2), 653–678. https://www.jstor.org/stable/40971343

Lazarus, R. S. (2013). *Fifty years of the research and theory of R. S. Lazarus: An analysis of historical and perennial issues*. Psychology Press.

Leshem, O. A. (2017). What you wish for is not what you expect: Measuring hope for peace during intractable conflicts. *International Journal of Intercultural Relations*, *60*, 60–66. https://doi.org/10.1016/j.ijintrel.2017.06.005

Leshem, O. A. (2019). The pivotal role of the enemy in inducing hope for peace. *Political Studies*, *67*(3), 693–711. https://doi.org/10.1177/0032321718797920

Leshem, O. A., & Halperin, E. (2020a). Hope during conflict. In S. C. van den Heuvel (Ed.), *Historical and multidisciplinary perspectives on hope* (pp. 179–196). Springer International. https://doi.org/10.1007/978-3-030-46489-9_10

Leshem, O. A., & Halperin, E. (2020b). Hoping for peace during protracted conflict: Citizens' hope is based on inaccurate appraisals of their adversary's hope for peace. *Journal of Conflict Resolution*, *64*(7–8), 1390–1417. https://doi.org/10.1177/0022002719896406

Leshem, O. A., & Halperin, E. (2021). Threatened by the worst but hoping for the best: Unraveling the relationship between threat, hope, and public opinion during conflict. *Political Behavior*. https://doi.org/10.1007/s11109-021-09729-3

Lopez, S. J., Snyder, C. R., & Pedrotti, J. (2003). Hope: Many definitions, many measures. In S. J. Lopez & C. R. Snyder (Eds.), *Positive psychological assessment: A handbook of models and measures* (pp. 91–107). American Psychological Association.

Menninger, K. (1959). The academic lecture: Hope. *American Journal of Psychiatry*, *116*(12), 481–491.

Miceli, M., & Castelfranchi, C. (2010). Hope: The power of wish and possibility. *Theory & Psychology*, *20*(2), 251–276. https://doi.org/10.1177/0959354309354393

Milona, M. (2020). Philosophy of hope. In S. C. van den Heuvel (Ed.), *Historical and multidisciplinary perspectives on hope* (pp. 99–116). Springer International. https://doi.org/10.1007/978-3-030-46489-9_6

Nesse, R. M. (1999). The evolution of hope and despair. *Social Research*, *66*(2), 429–469.

Nietzsche, F. (1878). *Human all too human*. Gordon.

Orne, M. (1968). On the nature of effective hope. *International Journal of Psychotherapy*, *5*(5), 403–410.

Robinson, M. D., & Clore, G. L. (2002a). Belief and feeling: Evidence for an accessibility model of emotional self-report. *Psychological Bulletin*, *128*(6), 934–960. https://doi.org/10.1037/0033-2909.128.6.934

Robinson, M. D., & Clore, G. L. (2002b). Episodic and semantic knowledge in emotional self-report: Evidence for two judgment processes. *Journal of Personality and Social Psychology*, *83*(1), 198–215. https://doi.org/10.1037/0022-3514.83.1.198

Roseman, I. J. (1991). Appraisal determinants of discrete emotions. *Cognition and Emotion*, *5*(3), 161–200. https://doi.org/10.1080/02699939108411034

Rosler, N., Cohen-Chen, S., & Halperin, E. (2017). The distinctive effects of empathy and hope in intractable conflicts. *Journal of Conflict Resolution*, *61*(1), 114–139. https://doi.org/10.1177/0022002715569772

Sagy, S., & Adwan, S. (2006). Hope in times of threat: The case of Israeli and Palestinian youth. *American Journal of Orthopsychiatry*, *76*(1), 128–133. https://doi.org/10.1037/0002-9432.76.1.128

Schoemaker, P. J. H. (1982). The expected utility model: Its variants, purposes, evidence and limitations. *Journal of Economic Literature*, *20*(2), 529–563.

Smith, C. A., & Ellsworth, P. C. (1985). Patterns of cognitive appraisal in emotion. *Journal of Personality and Social Psychology*, *48*(4), 813–838. https://doi.org/10.1037/0022-3514.48.4.813

Snyder, C. R. (1994). *The psychology of hope*. Free Press.

Snyder, C. R. (Ed.). (2000). *Handbook of hope: Theory, measures, and applications* (1st ed.). Academic Press.

Snyder, C. R., Cheavens, J., & Michael, S. T. (2005). *Hope theory: History and elaborated model* (J. Eliott, Ed.; pp. 101–118). Nova Science.

Spinoza, B. (1677/2000). *Ethics* (G. H. Parkingson, Trans.). Oxford University Press.

Staats, S. R. (1989). Hope: A comparison of two self-report measures for adults. *Journal of Personality Assessment*, *53*(2), 366–375.

Stone, R. A. (1982). *Social change in Israel, Attitudes and events*. Praeger.

Stotland, E. (1969). *The psychology of hope*. Jossey-Bass.

Telhami, S., & Kull, S. (2013). *Israeli and Palestinian public opinion on negotiating a final status peace agreement*. Saban Center at The Brookings Institution.

Tillich, P. (1965). *The right to hope*. Harvard Divinity School.
van Zomeren, M. (2021). Toward an integrative perspective on distinct positive emotions for political action: Analyzing, comparing, evaluating, and synthesizing three theoretical perspectives. *Political Psychology*, *42*(S1), 173–194. https://doi.org/10.1111/pops.12795
van Zomeren, M., Pauls, I. L., & Cohen-Chen, S. (2019). Is hope good for motivating collective action in the context of climate change? Differentiating hope's emotion- and problem-focused coping functions. *Global Environmental Change*, *58*, 101915. https://doi.org/10.1016/j.gloenvcha.2019.04.003
Wheatley, J. M. O. (1958). Wishing and hoping. *Analysis*, *18*(6), 121–131. https://doi.org/10.2307/3326568

4

The Determinants of Hope

We must accept finite disappointment but never lose infinite hope.

—Martin Luther King, Jr.

In the first week of January 2022, while COVID-19 was running wild in Israel-Palestine and elsewhere, I joined a group of Jewish Israelis for a four-day visit to the West Bank. The rare opportunity to tour Palestinian villages and cities in the Occupied Palestinian Territory (OPT) focused exclusively on the experiences of West Bankers and their daily plights and struggles. Seeing their dire situation was distressing for most Israeli participants, who came from various backgrounds but were mostly oblivious to the condition of West Bank Palestinians. Nearly all were shocked and bewildered by the devastating consequences of more than fifty years of Israeli military control.

In the evenings, the participants convened for a facilitated discussion to process the disturbing things they saw. Nir, a successful businessman and social entrepreneur, expressed genuine empathy toward the Palestinians. Though his wishes for ending the conflict seemed to be high, his expectations that resolution could be achieved seemed extremely low. "Seeing the scale of devastation, I don't know if there is any chance for transformation and peace. It's just too big, too entrenched, too intractable." Other participants expressed similar feelings of despair and hopelessness. After peeking into the heart of the conflict, its resolution seemed even more impossible.

What was most remarkable was the sharp contrast between the despair expressed by many Jewish Israeli participants and the hope expressed by the Palestinians we met. For example, Nabil from Jericho, who spent several years in Israeli prisons during the Second Intifada, was outspoken about his hope for change. "Of course peace will come! It's going to be hard, and it's going to take time, but there is simply no other option." Rawan, a Palestinian educator from Hebron, told us that teaching her high school students about peace and conflict transformation is her way of keeping her hope for peace alive. The fact that she has to wait hours at Israeli checkpoints to take her disabled son

Hope Amidst Conflict. Oded Adomi Leshem, Oxford University Press. © Oxford University Press 2024.
DOI: 10.1093/oso/9780197685303.003.0004

to hospital checkups did not seem to diminish her belief in the possibility of peace and resolution.

What makes one person hopeful and another hopeless about future peace? Is hope associated with the power relations in conflict? Is it correlated with age, religious observance, or other demographic and sociopolitical factors? Exploring the determinants of hope for peace is essential for at least two reasons. First, identifying the determinants of hope and hopelessness during conflict expands our theoretical understanding of the sociopsychological infrastructure of intractable conflicts. As noted, the theory of the sociopsychological infrastructure of intractability posits that conflicts become intractable due to a stable system of detrimental perceptions, emotions, beliefs, and attitudes held by members of rival societies (Bar-Tal, 2013; Coleman, 2003). According to the theory, the lack of hope for peace is one of these subjective constructs that maintain the conflict and propel its escalation (Rouhana & Bar-Tal, 1998). Understanding what generates hopelessness can thus improve our theoretical understanding of intractability.

The second reason to explore the precursors of hope and hopelessness is our urgent need to advance conflict resolution practice. Peacebuilders and peacemakers working in conflict zones often face skepticism and despair. They try to engage with policymakers, community leaders, and the masses to carve pathways toward peace. However, hopelessness among society members and their leaders often hampers these attempts (Kelman, 2018). Learning what jumpstarts hope can help societies mired in conflict escape the tragic cycles of violent disputes.

Given its importance, this chapter displays what we know about the predictors of hope for peace.[1] It is worth noting that the precursors of hope could be of different types and categories. Demographic factors, such as age, gender, income, and education, or fundamental worldviews like ideology and religious observance, might be associated with hope and hopelessness regarding future peace. In addition, peoples' subjective perceptions of the reality of conflict are also likely to impact their hopes for peace. One such perception concerns people's perceived threat from conflict escalation.

This chapter presents examples of the determinants of hope for peace from different categories. I begin by presenting existing work that identified several predictors of hope for peace. The rest of the chapter focuses on results from the Israeli-Palestinian Hope Map Project, a large-scale longitudinal

[1] Consistent with the bidimensional conceptualization of hope (Chapter 3), I use the word "hope" to refer to the two dimensions at the same time. However, when I want to refer to only one dimension I do not use "hope" but the name of the dimension, namely "wishes" or "expectations."

survey mentioned briefly in previous chapters. One of the aims of the Hope Map Project is to explore the determinants of hope for peace in conflict zones worldwide. The Hope Map Project utilizes the bidimensional model to conceptualize hope (i.e., separates wishes from expectations), making its results more relevant and accurate. I thus expand on the methods used in the Hope Map Project while also detailing how peace was defined. I then present the main findings regarding the predictors of hope for peace and end by offering critical insights from the project.

Existing Studies on the Determinants of Hope

We can learn much about the antecedents of hope from existing work that, intentionally or as a by-product, revealed its predictors. Halperin and colleagues (2008), for example, were interested in the correlation between hope and political ideology among Jewish Israelis. The researchers found that Jewish Israelis' self-identification as political hawks was negatively correlated to hope for peace ($r = -.33^{**}$).[2] The more people identified as hawks, the lower their levels of hope. Admittingly, we cannot know from this study if the term "hope" refers to peoples' wishes for peace or their expectations that peace will materialize.[3] However, the link between hope and ideological stance might be vital to understanding intractable conflicts and will be discussed later in this chapter.

In the same study, Halperin et al. also found that respondents' prior military service in the OPT was negatively associated with their hope for peace ($r = -.16^{*}$). Military service is mandatory in Israel, and many soldiers are sent to the OPT at some point in their three-year duty period. Most military activities in the OPT include frequent friction with Palestinian civilians and direct exposure to violence (Manekin, 2017). Serving long periods amidst violence and hostility might explain the low levels of hope for peace among Israeli soldiers. Yet again, we cannot be sure whether military service in the OPT is associated with low wishes for peace, low expectations for peace, or both. While it is plausible that the stressful experiences of soldiers stationed in the OPT would correlate with *low expectations* for peace, it can also be postulated that the same experiences will be associated *with greater wishes* to achieve an end to the conflict in peaceful ways (Leshem & Halperin, 2021).

[2] Note, throughout the book: †p < .1, *p < .05, **p < .01, p*** < .001. β refers to standardized coefficients.
[3] In this study, "peace" refers to peace with Arab nations and not only toward Palestinians.

Data I recently collected support the different effects of exposure to violence and the hope dimensions. Jewish Israelis who were directly exposed to conflict-related violence (witnessed or injured by conflict-related violence) have higher wishes for peace than those who were not exposed to direct violence ($\beta = .44^{*}$). Direct exposure to the devastation and pain caused by the conflict appears to go together with stronger desires to solve it peacefully. Interestingly, the expectation dimension was not associated with direct exposure to violence ($\beta = .08$). In sum, it seems that hope for peace is tied to people's direct encounters with the harsh reality of the conflict. The nuances of this tie, however, are still unclear.

Another line of work sought to identify the links between fundamental worldviews and hope for peace. One of these worldviews concerns people's beliefs about the malleability of humans and the world in general. Broadly speaking, people differ in their perceptions about whether humans, political phenomena, and the world in general are fixed or dynamic (Cohen-Chen et al., 2014; Dweck et al., 1995; Halperin et al., 2011). In a set of studies, Cohen-Chen and colleagues (2015) found a strong correlation between Jewish Israelis' belief that the world is continuously changing and their hope for peace ($r = .45^{**}$). To further demonstrate this connection, the researchers approached 70 Jewish Israeli train passengers and asked them to draw a quick sketch. Half were randomly assigned to sketch the world as dynamic and ever-changing, while the other half was requested to sketch the world as stable and fixed. Results from a survey administered later showed that those assigned to the "world is ever-changing" condition had higher hopes for peace than those who drew a picture of the world as fixed and rigid.

In this insightful experiment, Cohen-Chen et al. (2015) revealed a causal relationship between the belief in a changing world and Jewish Israelis' hope for peace. However, there are at least two limitations to what we can learn from this and other studies on hope during conflict. First, these studies' conceptualization of hope makes it impossible to distinguish between wishing for peace and expecting peace to come about. Another limitation of this and most studies on hope for peace is that it explored only one population in a dyadic conflict. Intergroup conflicts involve at least two parties, so focusing on one can only tell a partial story. The need to explore both parties to a conflict is particularly crucial in asymmetrical disputes in which the experiences of members from the lower-power party are fundamentally different from the experiences of the members of the side with superior power (Leshem & Halperin, 2020b; Thiessen & Darweish, 2018).

Shani and Boehnke (2017) addressed this gap by exploring hope among members of low- and high-power parties. In their study, the authors investigated hope for reconciliation among Jewish and Palestinian citizens of Israel who participated in a two-day intergroup encounter for youths.[4] As Shani and Boehnke describe, there are several types of models of intergroup encounters. One type of encounter focuses on the commonalities between the groups, while another model, particularly relevant to asymmetrical relations, stresses the power imbalance between the groups. A third model mixes the two approaches by dedicating some activities to intergroup commonalities and other activities to discuss intergroup disparities. The researchers showed that participating in a mixed-model encounter increased the hope for conciliation among Jewish Israelis and Palestinian citizens of Israel. By contrast, no increase in hope for reconciliation was observed in groups of Jewish and Palestinian youths who did not participate in the workshop. The authors conclude that intergroup encounters revolving around commonalities *and* disparities elicit positive outcomes such as increased hope for reconciliation among high-power and low-power participants. Again, because the bidimensionality of hope was overlooked, we cannot know whether increased hope reflects a rise in participants' desires for reconciliations or an increase in their belief that reconciliation is possible.

Power asymmetry is one of the central features of international conflicts (Klein et al., 2006) and is undoubtedly critical for understanding the conflict in Palestine-Israel (Rouhana, 2004). Because Palestinians and Israelis live in very different conditions of conflict, they are likely to have distinct aspirations regarding conflict resolution and future peace (Leshem & Halperin, 2020b). Addressing these distinct aspirations, Hasan-Aslih et al. (2020) sought to investigate West Bank Palestinians' hopes to end the occupation. The researchers showed that Palestinians' "motivation to feel hope" was predicted by the extent of their perceived setback in the Palestinian struggle for self-determination. The stronger the perceived setback, the higher the "motivation to feel hope" ($\beta = .22^*$). I want to offer an alternative interpretation of these results by adopting the bidimensional approach to hope. I suggest that the perceived setback simply predicted Palestinians' increased *wish* to end the occupation. The more Palestinians think the setbacks are substantial, the higher their desires to change the unjust situation of occupation and oppression.

Investigating the wishes and expectations of disfranchised groups (like Palestinians) is essential because it exposes the aspirations of those whose voices are least heard as well as their beliefs about the feasibility of actualizing

[4] Sometimes referred to as "Israeli Arabs," Palestinian citizens of Israel form about 20% of the Israeli citizenry.

these aspirations. Unfortunately, research on hope and other sociopsychological factors underpinning conflict among oppressed groups are scant (for exceptions, see Bar-Tal et al., 2017; Fink et al., 2022; Halperin et al., 2011; Sagy & Adwan, 2006; Shelef & Zeira, 2017). Even rarer are studies conducted simultaneously on high- and low-power societies locked in an asymmetrical conflict. One of the aims of the Hope Map Project is to address the lack of dyadic research on the sociopsychological infrastructure of conflicts. More specifically, the project seeks to unravel the determinants and outcomes of hope and hopelessness across conflict zones and contexts. Before we continue to look at the determinants of hope, I provide some information on the Hope Map Project.

The Hope Map Project

Background

The Hope Map Project is a global research project exploring the hopes for peace of people living in conflict zones. The project combines perspectives from political psychology, comparative politics, and international relations to reveal the role of hope (and hopelessness) in international conflicts. Three main objectives guide the project. The first is identifying the demographic, political, psychological, and situational antecedents of hope for peace across conflict contexts and stages. Does age have anything to do with the levels of hope for peace? Is religiosity a factor? And what is the relation between political ideology and hope? Answering such questions significantly contributes to our knowledge of how hope and hopelessness form and proliferate in times of conflict.

The second goal is to identify the direct consequences of hope for peace. This aspect of the project asks, "Does hope matter?" We can test, for example, whether hope for peace predicts people's support for peace-promoting policies or whether other factors are better predictors of support for or opposition to these policies. If hope predicts peace-promoting policies over and above other factors, there is a reason to invest in increasing hope in conflict zones worldwide. We can also explore other tentative outcomes of hope for peace, such as people's openness to the narrative of the outgroup or their voting preferences in national elections. Chapter 7 is dedicated to results and insights from this line of research.

The final goal is to offer a comprehensive comparison of hope for peace across conflict zones. Comparing the hope for peace in places such as Cyprus, Colombia, Israel-Palestine, and the Caucasus could teach us about the variations in the hopes for peace in different conflicts and the similarities across cultures and political circumstances. The Hope Map Project is an ongoing research initiative administered at set intervals in selected conflict zones. The project provides longitudinal insights into the dynamics of hope for peace across time and changing circumstances.

The Israeli-Palestinian Hope Map

The Israeli-Palestinian Hope Map examines the antecedents, outcomes, and dynamics of hope for peace in Israel, the West Bank, and the Gaza Strip. The first round was conducted in 2017, on a representative sample of 500 Jewish Israelis and 500 Palestinians from the West Bank and Gaza Strip.[5] In both societies, all respondents answered the same survey, which included measures of the two dimensions of hope for peace as well as tentative sociopsychological predictors and conflict-related outcomes. The survey also included demographic measures and a scale designed to gauge respondents' hawkish-dovish stance. Since 2017, two additional rounds of data collection were administered in the framework of the Israeli-Palestinian Hope Map Project (though in this book I report only on the data collected in the first round). As far as I know, the Hope Map Project offers the most comprehensive account to date of the hopes for peace of people embroiled in conflict.

The 500 Jewish Israeli participants of the 2017 Hope Map answered the survey online. However, online polling is not customary in the OPT due, in part, to Palestinians' suspicion that security forces are behind the data collection. As a result, Palestinian pollsters traditionally administer surveys face to face. To obtain a representative sample of Palestinian residents of the West Bank ($N = 300$) and the Gaza Strip ($N = 200$), enumerators followed a stratified probability sampling protocol, polling random households in cities, villages, and refugee camps across the OPT.

Defining Peace

The Hope Map Project investigates the hopes for peace of people mired in conflict. However, people interpret the word "peace" in different ways (Leshem & Halperin, 2020b; Malley-Morrison et al., 2013). If we want to explore people's

[5] Data were collected in only three days to minimize the potential influence of conflict-related events that might occur during data collection. The simultaneous collection of data among the two societies increased our ability to compare the hopes for peace of Palestinians and Israelis.

hope for "peace," we must be explicit about the type of "peace" we are referring to. At the same time, being too explicit and detailed has a significant drawback: it may exclude interpretations that might be the most relevant for certain people. Providing a very detailed definition of peace imposes a particular interpretation of what peace is. This interpretation might be irrelevant to many people, and so the task of capturing hopes for "peace" becomes ineffective.

The dilemma could be framed as a dilemma between flexibility and concreteness. Flexibility in the definition of peace allows for the inclusion of the different ways peace can be understood, which is essential if we want to gauge a wide gamut of perspectives. However, flexibility limits comparisons because some people might hope for one type of peace while others for a different type. My solution to this dilemma was to provide four definitions, or "types" of peace, ranging from flexible definitions to more concrete ones. I then measured Israelis' and Palestinians' *wishes* and *expectations* for each of the four types of peace. The definitions were presented in a particular order, from the most flexible to the most concrete.

The first definition was very flexible. In fact, it was not a definition per se because participants were asked to rate (on a scale from 1 to 6) how much they wished for "*peace as they define and understand it*" and how much they expect "*peace as they define and understand it*" to materialize. At first glance, using such an open-ended definition might seem pointless. However, precisely because this definition is open-ended, with no "strings attached," it has a unique advantage. The lack of specification offers the opportunity to tap into Palestinians' and Israelis' wishes and expectations for the *idea* of peace.

The subsequent definitions of peace were more concrete than the first but still maintained a generic flavor. The first one was "*a mutually agreed upon accord that ensures the interests of both peoples*," while the second definition of peace, "*a mutually agreed upon accord that ensures independence and freedom for Palestinians and security and safety for Israelis*," spelled out these interests. Note that the definitions are flexible enough to include many types of solutions that may cross participants' minds as long as these solutions are mutually accepted (for example, expelling the adversary to attain "peace" does not fall within the scope of these definitions). Although these definitions are general, they refer to peace as a reciprocal idea. At the stage of analysis, these two items collapsed well into one item (Wishes: ISR: $\alpha = .81$, PAL: $\alpha = .83$; Expectations: ISR: $\alpha = .9$, PAL: $\alpha = .77$) and so were used in the analyses as a single definition I term "reciprocal peace." Hope for reciprocal peace is a useful measure because it encapsulates people's wishes and expectations for a type of peace that includes reciprocity. Most of the analyses in this chapter focus on Israelis' and Palestinians' hope for reciprocal peace.

The last definition involved Israelis' and Palestinians' hope for a concrete definition of peace, specifically for the solution most debated inside and outside Israel-Palestine: the Two-State Solution (2SS). Speaking in very broad terms, the 2SS involves partitioning the land into two sovereign states, one for Jews and one for Palestinians. The 2SS includes several elements (like the status of Jerusalem, the future of the Palestinian refugees, and security issues). Yet the central premise of the 2SS is that the creation of an independent Palestinian state alongside Israel will bring an end to the parties' claims and terminate the violent dispute (Kelman, 2018). So far, all international attempts to resolve the conflict have been based on different versions of the 2SS.

The 2SS has enjoyed support from both societies in the first decade of the 21st century. However, this support has been dwindling ever since. In 2010, 71% of Israelis and 57% of Palestinians supported the 2SS. In 2020, the support declined to 42% among Israelis and 43% among Palestinians.[6] People's support for policies can be influenced by various factors, including the extent to which they think the policy is desirable and their assessment of its applicability (Leshem & Halperin, 2020c). Therefore, more than simply asking about support, the Hope Map Project seeks to reveal how much Israelis and Palestinians desire the 2SS and how much they think it feasible.

Results

Descriptive Results

As a start, it will be instructive to present the observed levels of hope for peace. I first note, perhaps not surprisingly, that the wishes for peace were high and the expectations were low across all definitions and the two societies. For instance, while the *wish* for reciprocal peace was above the scale's midpoint (4.12 among Palestinians and 4.62 among Jewish Israelis), the *expectations* for reciprocal peace were below the midpoint (2.58 among Palestinians and 2.23 among Jewish Israelis). Corresponding with previous research (Leshem, 2017; Sagy & Adwan, 2006), this difference between wishes and expectations is statistically significant (among Palestinians: $t = 19.4^{***}$, among Jewish Israelis: $t = 33.8^{***}$). The good news is that, across definitions, most

[6] Retrieved from the Palestinian/Israeli Pulse: https://www.pcpsr.org/en

Table 4.1 Israelis' and Palestinians' wishes for peace, by definition of peace

	PAL	ISR	Diff (PAL-ISR)
Open-Ended	4.78 (2.59)	5.05 (1.28)	t = −2.9**
Diff (OE-reciprocal peace)	t = 8.1***	t = 7.4***	
Reciprocal Peace	4.12 (1.61)	4.62 (1.43)	t = −5.27***
Diff (reciprocal peace-2SS)	t = 12.3***	T = 4.7***	
Two-State Solution	3.13 (1.89)	4.3 (1.75)	t = −10.2***

Note: All scales 1 to 6; standard deviations in parenthesis, *p < .05, **p < .01, p*** < .001.

Palestinians and Israelis want peace. The bad news is that most think it can never happen.

Means and standard deviations of participants' hope for peace are provided in Tables 4.1 and 4.2. The tables also include the results from within-subject comparisons revealing the differences between the hope for the different definitions. In addition, I present results from between-sample t-tests comparing the hope of Israelis and Palestinians.

First, looking at Israelis' and Palestinians' wishes for the different definitions of peace (Table 4.1), it appears that, in both societies, the wish for peace dropped as the definitions became more concrete (see also Leshem, 2017, for a similar finding). One way to explain this decline is that it might be relatively easy, even obvious, to desire the idea of peace. However, when the definitions become less abstract, the perceiver must think about peace in more practical terms, such as reciprocating or making concessions. When elements like reciprocation and the need to compromise become tied to peace, peace becomes less desirable.

Another interesting point is that although the wish for peace among Palestinians was relatively high, it was significantly lower than among Israelis in all three definitions. This observation holds in a simple t-test comparison. A more thorough analysis shows that the differences disappear after accounting for demographic and sociopolitical factors (Leshem & Halperin, 2020a). Stated otherwise, when the impact of demographic and sociopolitical factors is accounted for, the wish for peace among Palestinians and Israelis is very much the same.

Looking at Palestinians' and Israelis' expectations for the different definitions of peace (Table 4.2) reveals mixed findings. For example, Israelis' expectation for the open-ended definition of peace and reciprocal peace was similar, but there was a significant difference between the two among Palestinians. At the same time, Israelis' expectation for the 2SS was higher

Table 4.2 Israelis' and Palestinians' expectations for peace, by definition of peace

	PAL	ISR	Diff (PAL-ISR)
Open-Ended	2.84 (1.7)	2.29 (1.27)	t = 5.77***
Diff (OE-reciprocal peace)	t = 3.8***	t = 1.3^{ns}	
Reciprocal Peace	2.58 (1.4)	2.23 (1.21)	t = 4.2***
Diff (reciprocal peace-2SS)	t = 0.9^{ns}	T = −7.16***	
Two State Solution	2.65 (1.67)	2.56 (1.38)	t = 0.9^{ns}

Note: All scales 1 to 6; standard deviations in parenthesis, *p < .05, **p < .01, p*** < .001.

than for reciprocal peace, but there was no difference in the expectations for the two definitions among Palestinians. In sum, unlike the evident decline in people's *wish* for peace when peace turns from flexible to concrete, it is hard to find consistent patterns in Israelis' and Palestinians' expectations for peace across definitions.

Perhaps most surprising is the difference between Israelis' and Palestinians' expectations for each definition. Though both Israelis and Palestinians believed that peace is unlikely, Palestinians had higher expectations that peace would transpire (far-right column), at least for the first two definitions. This pattern might seem counterintuitive because, if anything, the harsh conditions in the OPT are likely to make Palestinians skeptical rather than optimistic about the chances of peace. I present a tentative explanation for this finding in the discussion section.

Determinants of Hope

We now turn to the precursors of hope for peace revealed by the Israeli-Palestinian Hope Map. To identify the precursors of hope, participants' demographics (income, levels of education, gender, and age) and general worldviews (religious observance, hawkish-dovish stance, political efficacy, and acceptance of uncertainty) were entered as tentative predictors in two regression models, one predicting participants' wishes for peace and the other their expectations that peace will occur. For the sake of brevity, I focus here on the predictors of Israelis' and Palestinians' hope for reciprocal peace (though trends were similar for the other definitions). I chose to focus on hope for reciprocal peace because it captures people's wishes and expectations for a solution that respects the interest of both peoples (security for Israelis and independence for Palestinians) without imposing a specific configuration of the solution. In this chapter, I zoom in on several predictors that I found most interesting.[7]

Age. Does age predict the hopes for peace of those mired in intractable conflict? And, if so, in which direction? Two opposite hypotheses can be suggested. The first is that those who lived for an extended period amidst conflict will have lower wishes and expectations for peace than the young. Compared to the young, the old might be weary of wishing for peace, burnt out, and exhausted from maintaining a desire that was not met. Witnessing many years of war and futile attempts to negotiate a deal are also likely to lower the expectations for peace among older citizens. Hope for peace is thus likely to be lower among older generations compared to the young. This assumption is corroborated by research showing that older people have a more challenging time imagining future events (Addis et al., 2008), so imagining peace might be more difficult for the older cohorts.

On the other hand, past research has shown that age is associated with better conflict resolution skills (Grossmann et al., 2010) and that older people try to shift attention to positive rather than negative mindsets (Carstensen et al., 1999). In addition, the broader perspectives and richer life experiences that older people possess might be an asset for maintaining hope. Broad perspectives are a rarity among young Palestinians and Israelis exposed to cynical and corrupt politics in the past two decades. This exposure might lead to a very skeptical stance among the young and perhaps more indifference to the notion of peace. From the narrow perspective of the young, peace is neither desirable nor possible.

Analysis shows that the second hypothesis fits the data more accurately. The older people are, the higher their wishes ($\beta = .12^{***}$) and expectations for peace ($\beta = .09^{**}$). Looking separately at each population, some nuances are revealed. The effect of age on the wish for peace was driven by the Jewish Israeli sample, whereas the Palestinian sample drove the effect of age on expectations. That said, the direction of the effects was positive in both samples, indicating that the older generations in both societies have more hope for peace than the young on both dimensions.

A recent study by Hasler, Leshem, and colleagues validates the link between older age and hope for peace using novel methods (Hasler et al., 2023). First, analyzing almost three hundred public opinion polls conducted in Israel between 1994 and 2017, we revealed a consistent pattern

[7] Respondents' income and levels of education did not predict their hope for peace. Gender, however, did, such that women had higher wishes and expectations for peace compared to men. For further reading see Leshem and Halperin (2020a).

demonstrating that the older cohorts have higher hope for peace than younger cohorts (measured here only on the expectation dimension). Based on these findings, Hasler wanted to test whether making young people *feel* old increases their hope for peace. To achieve this, about 150 Jewish Israeli students were asked to participate in a virtual reality (VR) study. Half were randomly assigned to experience a VR aging simulation designed especially for the study. In the simulation, participants saw themselves as 80-year-old individuals (men or women, depending on the participant's gender). Thanks to VR technologies, the simulation created a holistic immersive experience. When participants moved about, their "older selves" moved in complete accordance. After the immersive experience, participants filled out a survey with questions about their hope for peace (in this case, the expectation dimension).

We show that young students who saw themselves as an older person had higher hope for peace than students in the control group who were immersed in a body of a 25-year-old. Put simply, feeling like an 80-year-old increased young Jewish Israelis' belief in the possibility of peace. Overall, age impacts how we think and consequently affects our wishes and expectations for peace. The implications of the link between age and hope are discussed at the end of this chapter.

Religiosity. Hope is advocated in the scriptures and teachings of many religions (Elliot, 2020), especially in monotheistic religions, where hoping for salvation is a fundamental aspect of believing in God's power and grace (Brunner, 1956). Are the observant indeed more hopeful? Sethi and Seligman (1993) used self-reported scales to show that Christians, Muslims, and Jews affiliated with fundamental denominations are more optimistic than Christians, Muslims, and Jews affiliated with moderate or liberal denominations. One way to explain these findings concerns the secular propensity to doubt and question. Perhaps the absolute reliance on God's grace provides believers with a strong sense of hope and optimism. The secular, in contrast, must always stay somewhat in the dark, with less conviction about how the future will unfold.

Based on this logic, we should expect religious observance to be positively associated with hope for peace in Palestine-Israel. The more devout one is, the higher one's wishes and expectations for peace in the Holy Land. Nevertheless, results from the Hope Map Project reveal an opposite trend. In both societies, religiosity was one of the most robust predictors of low wishes (PAL: $\beta = -.13^{**}$, ISR: -29^{***}) and expectations (PAL: $\beta = -.12^{**}$; ISR: $\beta = -.1^{*}$) for peace, even after accounting for the impact of ideology. The data show that the more observant one is, the lower one's desires and expectations for peace in the Holy

Land. These findings might seem unsurprising, given that the conflict has a profound religious component (Canetti et al., 2019). Nevertheless, the fact that the impact of religiosity remained even after controlling for participants' hawkish-dovish stance shows how weighty the religious factor is. For example, identifying as ultra-orthodox or religious (compared to "traditional" or "secular") was the strongest predictor of low wishes for peace among Jewish Israelis. Suggested lessons from these findings for research and peacebuilding practice are presented in the discussion.

Hawkish-dovish stance. By its very nature, people's inclination to adopt a hawkish versus dovish stance toward conflict is strongly related to a host of conflict-related attitudes and behaviors. Past studies have shown that Israelis identifying as right-wing are more inclined to dehumanize the adversary (Adler et al., 2022; Maoz & McCauley, 2008), oppose compromise with the rival party (Shulman et al., 2021), and are more reluctant to support peacebuilding initiatives (Leshem, 2019). It thus comes as no surprise that, across both societies, hawks had lower wishes and expectations for reciprocal peace compared to doves (wish β: = −.14***, expectations: −.21***). But does the hawkish-dovish stance predict the tendency to hope for the *general notion of peace*? Are the wishes and expectations for peace, *any peace*, dependent on one's political ideology?

This question is essential for conflict scholars, political scientists, and others working on the connection between ideology and international conflicts. It is important because a common assumption about leftists' and rightists' approaches toward peace is that the differences concern practical considerations of peace and the ways to get there (Galtung, 1969; Lupovici, 2013). The general presumption is that *everybody* wants peace and that differences between left and right lie in how peace should be configured and implemented. The debate between leftists and rightists, so they say, is about content, process, and enforcement, not about the mere desirability of the idea of peace.

I set to explore this question by testing whether the hawkish-dovish stance predicts Israelis' and Palestinians' *wish* for the most flexible definition of peace, "peace as you defined and understand it." If the hawkish-dovish stance predicts people's wish for this entirely open-ended description, then the debate between left and right is also about the desirability of the abstract idea of "peace." Results reveal that Palestinians' wish for open-ended peace was not associated with their general hawkish-dovish stance. Stated differently, Palestinian hardliners and doves equally think that the idea of peace is desirable. However, among Jewish Israelis, the picture is different. Doves' wish for the notion of peace was much higher than hawks' (β = .16***). It seems that, in Israel, the mere desire for the abstract idea of peace hinges on ideology.[8]

Political efficacy beliefs. people's belief in their ability to influence political processes is a well-studied topic in political science (e.g., Balch, 1974; Nijs et al., 2019). The scope of this chapter limits the ability to discuss the different aspects of political efficacy beliefs. I will thus only refer to the narrow sense of the term that concerns the degree to which people believe they can impact politics (Sulitzeanu-Kenan & Halperin, 2013). Generally speaking, those scoring high on political efficacy think that the public has a decisive role in politics, that public opinion matters, and that society members have some influence on political processes. People scoring low on political efficacy believe that ordinary people are not part of the political game but are merely pawns in a grander game played by elites. People with low-efficacy beliefs think that no matter what they do, citizens cannot shape politics.

Do political efficacy beliefs predict Israelis' and Palestinians' hope for peace? If so, which dimension is more susceptible to political efficacy? Let us look at each dimension separately. The wish for peace (i.e., the level of desires and aspiration for peace) is not likely to be tied to the belief that people are or are not efficacious political actors. One could have strong desires for peace without believing that the public can bring about this desired change. People's expectations for peace, on the other hand, might be strongly tied to political efficacy beliefs, especially if peace is understood as having a bottom-up component. Expectation, but not wishes for peace, was thus hypothesized to be positively associated with political efficacy beliefs.

Results confirm this hypothesis. Political efficacy beliefs predicted Jewish Israelis' ($\beta = .15^{**}$) and Palestinians' ($\beta = .08^{\dagger}$) expectations for peace but not their wishes (PAL: $\beta = .05^{ns}$, ISR: $\beta = .02^{ns}$). This finding may point to the connection between bottom-up social change and intergroup peace. When people believe in their ability to shape their political future, they are more inclined to think peace is possible. This finding also points to the utility of the bidimensional model. It shows that factors can be associated with one dimension of hope but independent from the other. The utility of separating the dimensions is also demonstrated in the connection between hope and uncertainty.

Acceptance of uncertainty. People differ in their approach to uncertainty (Fiske, 2010; Freeston et al., 1994). Some people have a hard time with the idea

[8] Looking at the expectations that peace (in its open-ended form) will materialize shows a consistent pattern. In both populations, a hawkish stance is the strongest predictor of low expectations for peace (PAL: $\beta = -.20^{***}$, ISR: $\beta = -.35^{***}$). It seems that hardliners in both societies are skeptical of achieving peace, any peace.

that the future is unpredictable, whereas others are more at ease with the fact that not much is known about the future. For people living in a century-old conflict, the reality of conflict, violence, and hostility is tragically familiar and predictable. Peace, on the other hand, is unfamiliar and requires taking risks for an uncertain future. Thus, accepting the unpredictable nature of the future might be associated with Israelis' and Palestinians' hope for peace. But with which hope dimension: wishes, expectations, or both?

Examining each dimension separately, I postulated that participants' expectations for peace will not be associated with their acceptance of uncertainty. After all, people could feel entirely comfortable with uncertainty and have varying degrees of expectations for peace. However, the extent to which people desire peace (the wish dimension) could, at least in potential, be correlated with how comfortable they are with the unpredictability of the future. Those uncomfortable with uncertainty might suppress their desire for peace, which is unfamiliar and unpredictable. By the same token, people who are more at ease with uncertainty might allow themselves to wish for unfamiliar and unpredictable peace.

These assumptions were confirmed in both populations. The degree to which Palestinians and Jewish Israelis accept uncertainty did not predict their expectations for peace but did predict their wishes for peace. The greater the acceptance, the higher the wish (PAL: $\beta = .08\dagger$, ISR: $\beta = .08\dagger$) though the effects were only marginal. Recent data collected among 600 Israeli Jews provide additional support for these findings but from a different angle.

The theory of the Big-Five personality traits is one of the most cited in personality psychology (Cobb-Clark & Schurer, 2012) and is often used by political psychologists who study the connection between political behavior and stable personality dispositions (Gerber et al., 2011). Together with students, I sought to investigate the connection between the Big-Five traits (Openness to Experience, Conscientiousness, Extraversion, Agreeableness, and Neuroticism) and hope for peace. Building on the observed connection between acceptance of uncertainty and the wish for peace, we hypothesized that Openness to Experience (i.e., people's disposition to welcome new experiences) would predict their wishes but not their expectations for peace.

The data fully supported this hypothesis. Openness to Experience predicted Jewish Israelis' wish for peace even after accounting for potential influences of other traits, demographic measures, and sociopolitical factors like political ideology. In fact, out of the Big Five, Openness to Experience was the only personality disposition associated with the wish for peace. No personality trait, including Openness to Experience, was associated with the expectations for peace. Indeed, one of the most tragic aspects of intractable conflicts is that they become predictable. People locked in protracted disputes are well aware of the dangers and pain brought about by conflict. They are not surprised when conflict escalates or when war erupts. This familiarity is detrimental because it makes people prefer the destructive but predictable reality of conflict to the uncertain and unfamiliar reality of peace.

Threat perceptions. So far, this chapter has focused on predictors that are not necessarily intrinsic to the reality of conflicts. Though they substantially influence people's hope for peace, factors like age, political efficacy, acceptance of uncertainty, and even people's hawkish-dovish stance are not particular to conflicts. In this subsection, I turn to examine the link between hope and an experience unique to life in a conflict zone: people's threat from escalation. *Threat* is a deep sense of vulnerability to future harm that may result in a loss (Gilbert, 2005). In conflict, people's threat perceptions revolve around their concern about the recurrence of violence and hostilities. People involved in violent conflicts imagine bombs, missiles, night raids, the cries of terrified children, and the unbearable devastation that comes when a loved one is killed. While conflict escalation is an undesired turn of events, it is a possible, even probable, scenario.

Threat perception refers to peoples' beliefs about the possibility and severity of *undesirable* outcomes. Like hope, threat perceptions comprise two dimensions (Leshem & Halperin, 2021). The first is the perceived likelihood of an adverse event. The more we believe that an adverse event is probable, the higher the perceived threat. During conflict, the perceived threat is high when escalation is believed to be probable and low when escalation is believed to be unlikely. However, the perceived threat is also based on assessing the *severity* of the adverse event. The more the adverse event is perceived as severe (i.e., harmful, destructive), the higher the perceived threat. Together, the two components of threat (perceived likelihood and perceived severity) constitute the total perceived threat. The question is, how are they linked to hope?

I start by examining the perceived *severity* of escalation. During conflict, people might differ in the extent they believe future escalation will have severe consequences. Some might perceive escalation as entailing unbearable suffering and damage, whereas others may be less extreme in their perceptions about the severity of future escalation. It could be argued that those who envision future violence as devastating and extremely unbearable would be more eager to end the conflict and hasten peace than those who perceive future escalation in less intense terms. I thus hypothesized that, all else equal, the more one's threat stems from perceiving escalation as *severe* and destructive, the more one would *wish* for peace.

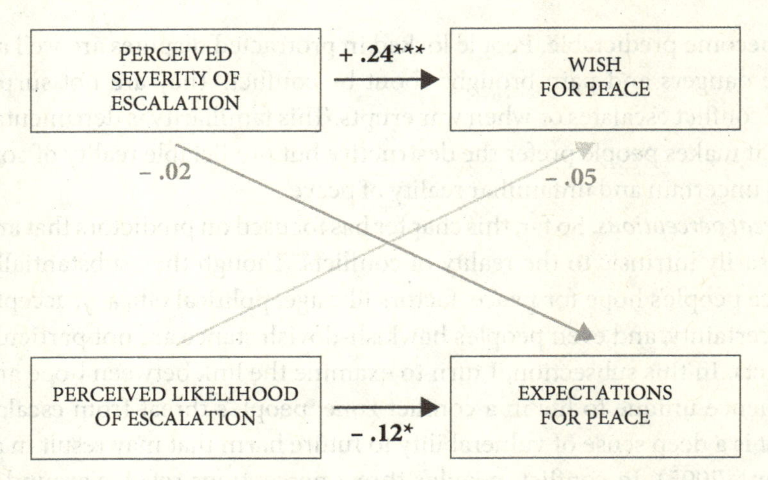

Figure 4.1 Relations between threat components and hope dimensions.

Next, people involved in conflict might also differ in their perceptions of the *likelihood* of escalation. Some might think that future escalation is probable, even inevitable, while others may think that escalation is not so likely. All else equal, those who think that future violence is *likely* are more inclined to have low *expectations* for peace.[9] In sum, hope's link with perceived threat is nuanced. The part of threat that derives from perceiving future violence as *severe* is postulated to be positively associated with their *wish* for peace, while the part of threat that stems from perceiving violence as *likely* is hypothesized to be negatively associated with the *expectations* for peace. Do data from Israel-Palestine support these hypotheses?

As shown in Figure 4.1, results confirm the postulated connections between perceived threat and hope for peace.[10] Perceiving future escalation as severe predicted Israelis' and Palestinians' wishes for peace ($\beta = .24^{***}$), while perceiving future escalation as likely predicted Israelis' and Palestinians' low expectations for peace ($\beta = .12^{*}$). Interestingly no significant crossover effects were detected.[11]

[9] This connection might stem from the tendency to connect the frequency of conflict-related events with the improbability of peace, though frequent violent confrontations often precede successful peace processes (Goertz & Diehl, 1995).

[10] To test the association between threat and hope, two new regression models were estimated, one predicting the wish and the other the expectations for reciprocal peace. In both models, the two threat components were the key tentative predictors while controlling for demographic measures, religiosity, and hawkish-dovish stance. Interaction terms between the threat components and nationality were also added to test whether trends were similar in both societies. Results show that the interactions were not significant and that results replicate in each society.

[11] Results shown here are pooled from the 500 Jewish Israelis and 300 Palestinians from the West Bank. Unsurprisingly, perceived threats of Palestinians from Gaza were extremely high on both dimensions. The highly skewed and invariant reports of threat from Gaza made it impossible to include them in the analyses.

What do these results reveal about the relationship between threat perceptions, hope, and conflict? One valuable insight may shed light on why protracted conflicts, though characterized by fluctuating levels of violence, are exceptionally resilient to resolution. People's perceived threats will likely rise on both dimensions when conflict escalates. Heightened threat on the severity dimension increases people's wish for peace and may consequently raise their support for negotiations and conciliatory policies. At the same time, increased threat derived from the perceived likelihood of violence decreases expectations that peace can materialize and, in turn, decreases support for the same policies. In short, the opposite impacts of the two dimensions of threat entirely or partially cancel each other. During relative calm, both threat components drop, and so peace might seem more feasible but, at the same time, less desirable (compared to when violence is surging). Again, the opposite trends generate a standstill concerning public support for peacebuilding and reconciliation.

Discussion and Conclusion

I began this chapter with a trip to the West Bank and the opposite outlooks of Nir and Nabil. Nir, the Israeli participant, was hopeless about the chances for peace after witnessing the harsh reality in the West Bank. Nabil, who has lived within this reality since childhood, expressed hope and optimism regarding future peace. How could these differences be explained? Of course, one way is to interpret these differences as dispositional. Nir could be a born skeptic, and Nabil an optimist by nature. Though personality differences are plausible in this anecdotal example, I would like to devote most of the discussion to a broader political interpretation of Nabil's and Nir's opposite perspectives. This interpretation also explains why Palestinians' expectations for peace were significantly higher than Israelis' in the survey.

A right-wing shift in the politics of the Israeli leadership toward peace characterizes the past decades. Until the collapse of the Camp David Summit in 2000 and the Second Intifada that erupted in its aftermath, peace with the Palestinians seemed like a huge challenge. However, despite the challenges, the main message conveyed to the Israeli public by its leaders was that the striving for peace is worthy. However, during Netanyahu's twelve-year reign, this hopeful message gradually transformed into what Oakeshott (1996) termed the "politics of skepticism" (Navot et al., 2017). In its purest form, political skepticism seeks to reduce the government's responsibility to improve citizens' quality of life by encouraging skepticism about the possibility of grand positive change. When citizens are convinced that the future holds no big triumphs, their expectation from their government diminishes, which reduces public demand for reforms and changes. Internalizing this skepticism makes citizens docile, creating additional leeway for the government to implement its will. Navot et al. (2017) argue that Netanyahu's success can be attributed to his ability to spread the idea of skepticism and turn it into political power.

As previously explained, skepticism has many psychological advantages. It protects people from the deep despair caused by dashed hope and exonerates them from commitment and responsibility. Therefore, it is not surprising that skeptical leadership can be attractive to constituents. Skeptic leaders seem less naïve and more realistic. They encourage citizens to focus on the present and avoid delusional dreams. In the context of the Palestinian-Israeli conflict, the politics of skepticism is best exemplified in Netanyahu's famous declaration that Israelis will have to learn to live on their swords forever (Haaretz, October 25, 2015). Naftali Bennett, Netanyahu's successor, fully adopted this approach, claiming peace is impossible at this stage.[12] Bennett's skeptical outlook may sound like a straightforward, nonideological prognosis. However, it hides an ideological message: "peace is neither urgent nor feasible."

Much can be said about the political processes in Palestinian society in the past decade. In their competition against Israel, the Palestinians seemed to lose on every front. Their diplomatic effort to garner support from the Arab world proved as futile as their attempts to pressure the West to hold Israel accountable for its policies in the OPT. While thousands of Palestinians lost their lives in military campaigns led by Israel in Gaza and numerous incidents in the West Bank, Israel increased its civilian presence in the West Bank and signed peace treaties with the United Arab Emirates (UAE) and Bahrain.

Despite these calamities, Palestinian officials continued to stress their belief in the desirability and possibility of peace. "Despite the historical injustice that has been inflicted upon our people, their (Palestinians') desire to achieve a just peace . . . will not diminish," declared Palestinian President Abu-Mazen in 2010. Ten years later, at the United Nations Security Council, Abu Mazen repeated his hopeful stance. "I have come before you today to say that peace between Israelis and Palestinians is still possible. It's achievable. I have come to build an international partnership toward peace." Similar declarations were made public by the Palestinian Committee for the Interaction with Israeli

[12] .J Lis (August 30, 2021), "No diplomatic process with the Palestinians," source close to Bennett says after Gantz meets Abbas. Haaretz. https://www.haaretz.com/israel-news/.premium-gantz-meets-with-abbas-in-ramallah-to-discuss-security-economy-1.10163773

Society.[13] In its formal publication, this official committee stresses the urgency and the feasibility of reaching a viable solution in the form of a two-state compromise.

Is Palestinian optimism only a show intended for the non-Palestinian audience? Are Israelis more "realistic"? My perspective is that the difference between Israelis and Palestinians in their expectations for peace results from their distinct experience of the conflict. Simply put, for the Palestinians and their leaders, pessimism is not an option. The difficult conditions of life under occupation require tremendous amounts of hope on both dimensions. Indeed, the struggle for self-determination and justice necessitates an unbreakable belief in the possibility of peace and its assumed results—freedom and political independence. On the other hand, Jewish Israelis, who are living under more benign circumstances, can "afford" to be more skeptical about the chances for peace. Believing that peace is possible is currently not a necessity for them. These different approaches are reflected in the outlooks of Nabil and Nir and the study results presented in this chapter. As Chapter 6 reveals, this gap is also apparent when looking into Palestinians' and Israelis' expressions of hope for peace in speeches given to international audiences.

I now wish to draw attention to the applied implications of some of the findings, specifically about the association between hope, age, and religiosity. With some variations, results show that the young and the religious have lower wishes and expectations for peace than the old and secular. What does this mean for conflict resolution in Israel-Palestine? How do peacebuilders instill hope for peace where most of the population is young and traditional (Israeli Central Bureau of Statistics, 2018; Palestinian Central Bureau of Statistics, 2017)?

I speculate that the answer to this challenging puzzle might not lie so much within the concept of "hope" but instead with the concept of "peace." The failure of the Oslo Accords and other processes that shaped Palestinian-Israeli relations in the past two decades has severely damaged the reputation of "peace" (Kaufman & Freitekh, 2021). "Peace" has stepped down from its status as an inspirational ideal and turned into a hollow concept with the help of an ineffective peace movement, persuasive right-wing rhetoric, and infamous initiatives like the "Trump Peace Plan." Peace, as a brand, has lost its power.

[13] This formal committee established by the Palestine Liberation Organization (PLO) seeks to create a direct channel of communication between the Palestinian leadership and the Israeli public. Through social media and interpersonal interaction, the committee exposes the formal positions of the Palestinian Authority to the general Israeli public and creates platforms for dialogue.

The scope of this book cannot cover the breadth of the conversation concerning the (real or imagined) demise of the idea of peace in Israel-Palestine (for a more extensive debate, see Morris, 2009; O'Malley, 2016; Tilley, 2010). However, what becomes gradually apparent to those who wish to promote peace in the region is that "peace" must be rebranded to survive. The findings presented in this chapter suggest that this rebranding must address the needs of those whose hope for peace is the lowest, namely, the young and the religious. Scholars and practitioners would benefit from investigating how peace can attract the young. Infusing the notion of peace with ideas like innovation and entrepreneurship is perhaps one way peace can connect with younger generations. Rethinking how to make peace compelling for the religious sectors in both societies is also necessary (Canetti et al., 2019). One avenue to explore is whether the reconciliation approach is more appropriate for religious people than the policy-oriented mindset.

Implementing these ideas is extremely difficult. Even more challenging is the need to address the negative association between hope for peace and hawkish political ideology. As shown, the desires and expectations for peace are lower among political hawks even when the definition of peace is open for interpretation. This hints at the more profound differences across the political spectrum about the mere desirability of the idea of peace. Perhaps something could be learned from exploring the desires and expectations for peace of peacebuilders, women and men who work tirelessly to advance peace and conciliation even during times of escalations. The subsequent chapter journeys to the hearts and minds of Palestinian and Israeli peace activists to help shed light on how hope functions at the frontlines of peacebuilding work.

References

Addis, D. R., Wong, A. T., & Schacter, D. L. (2008). Age-related changes in the episodic simulation of future events. *Psychological Science, 19*(1), 33–41. https://doi.org/10.1111/j.1467-9280.2008.02043.x

Adler, E., Hebel-Sela, S., Leshem, O. A., Levy, J., & Halperin, E. (2022). A social virus: Intergroup dehumanization and unwillingness to aid amidst COVID-19. Who are the main targets?☆. *International Journal of Intercultural Relations, 86*, 109–121. https://doi.org/10.1016/j.ijintrel.2021.11.006

Balch, G. I. (1974). Multiple indicators in survey research: The concept "sense of political efficacy." *Political Methodology, 1*(2), 1–43.

Bar-Tal, D. (2013). *Intractable conflicts*. Cambridge University Press.

Bar-Tal, D., Diamond, A. H., & Nasie, M. (2017). Political socialization of young children in intractable conflicts: Conception and evidence. *International Journal of Behavioral Development, 41*(3), 415–425. https://doi.org/10.1177/0165025416652508

Brunner, E. (1956). *Faith, Hope, and Love*. Philadelphia: The Westminster Press.

Canetti, D., Khatib, I., Rubin, A., & Wayne, C. (2019). Framing and fighting: The impact of conflict frames on political attitudes. *Journal of Peace Research*, 002234331982632. https://doi.org/10.1177/0022343319826324

Carstensen, L. L., Isaacowitz, D. M., & Charles, S. T. (1999). Taking time seriously: A theory of socioemotional selectivity. *American Psychologist*, *54*(3), 165. https://doi.org/10.1037/0003-066X.54.3.165

Cobb-Clark, D. A., & Schurer, S. (2012). The stability of big-five personality traits. *Economics Letters*, *115*(1), 11–15. https://doi.org/10.1016/j.econlet.2011.11.015

Cohen-Chen, S., Crisp, R. J., & Halperin, E. (2015). Perceptions of a changing world induce hope and promote peace in intractable conflicts. *Personality and Social Psychology Bulletin*, *41*(4), 498–512. https://doi.org/10.1177/0146167215573210

Cohen-Chen, S., Halperin, E., Crisp, R. J., & Gross, J. J. (2014). Hope in the Middle East: Malleability beliefs, hope, and the willingness to compromise for peace. *Social Psychological and Personality Science*, *5*(1), 67–75. https://doi.org/10.1177/1948550613484499

Coleman, P. T. (2003). Characteristics of protracted, intractable conflict: Toward the development of a metaframework-I. *Peace and Conflict: Journal of Peace Psychology*, *9*(1), 1–37.

Dweck, C. S., Chiu, C., & Hong, Y. (1995). Implicit theories and their role in judgments and reactions: A word from two perspectives. *Psychological Inquiry*, *6*(4), 267–285. https://doi.org/10.1207/s15327965pli0604_1

Elliot, D. (2020). Hope in theology. In S. C. van den Heuvel (Eds.), *Historical and multidisciplinary perspectives on hope* (pp. 117–136). Springer International. https://doi.org/10.1007/978-3-030-46489-9_7

Fink, O., Leshem, O. A., & Halperin, E. (2022). Oppression and resistance: Uncovering the relations between anger, humiliation and violent collective action in asymmetric intergroup conflict. *Dynamics of Asymmetric Conflict*, *15*(3), 1–14. https://doi.org/10.1080/17467586.2022.2112408

Fiske, S. (2010). *Social beings* (2nd ed.). Wiley.

Freeston, M. H., Rhéaume, J., Letarte, H., Dugas, M. J., & Ladouceur, R. (1994). Why do people worry? *Personality and Individual Differences*, *17*(6), 791–802. https://doi.org/10.1016/0191-8869(94)90048-5

Galtung, J. (1969). Violence, peace, and peace research. *Journal of Peace Research*, *6*(3), 167–191.

Gerber, A. S., Huber, G. A., Doherty, D., & Dowling, C. M. (2011). The Big Five personality traits in the political arena. *Annual Review of Political Science*, *14*(1), 265–287. https://doi.org/10.1146/annurev-polisci-051010-111659

Gilbert, C. G. (2005). Unbundling the structure of inertia: Resource versus routine rigidity. *Academy of Management Journal*, *48*(5), 741–763. https://doi.org/10.2307/20159695

Goertz, G., & Diehl, P. F. (1995). The initiation and termination of enduring rivalries: The impact of political shocks. *American Journal of Political Science*, *39*(1), 30–52. https://doi.org/10.2307/2111756

Grossmann, I., Na, J., Varnum, M. E. W., Park, D. C., Kitayama, S., & Nisbett, R. E. (2010). Reasoning about social conflicts improves into old age. *Proceedings of the National Academy of Sciences*, *107*(16), 7246–7250. https://doi.org/10.1073/pnas.1001715107

Halperin, E., Bar-Tal, D., Nets-Zehngut, R., & Drori, E. (2008). Emotions in conflict: Correlates of fear and hope in the Israeli-Jewish society. *Peace and Conflict: Journal of Peace Psychology*, *14*(3), 233–258. https://doi.org/10.1080/10781910802229157

Halperin, E., Russell, A. G., Trzesniewski, K. H., Gross, J. J., & Dweck, C. S. (2011). Promoting the Middle East peace process by changing beliefs about group malleability. *Science*, *333*(6050), 1767–1769. https://doi.org/10.1126/science.1202925

Hasan-Aslih, S., Shuman, E., Goldenberg, A., Pliskin, R., van Zomeren, M., & Halperin, E. (2020). The quest for hope: Disadvantaged group members can fulfill their desire to feel hope, but only when they believe in their power. *Social Psychological and Personality Science, 11*(7), 879–888. https://doi.org/10.1177/1948550619898321

Hasler, B. S., Leshem, O. A., Hasson, Y., Landau, D. H., Krayem, Y., Blatansky, C., Baratz, G., Friedman, D., Psaltis, C., Cakal, H., Cohen-Chen, S., & Halperin, E. (2023). Young generations' hopelessness perpetuates long-term conflicts. *Scientific Reports, 13*(1), 4926. https://doi.org/10.1038/s41598-023-31667-9

Israeli Central Bureau of Statistics. (2018). Israel's Statistical Abstract 2018. https://www.cbs.gov.il/he/publications/Pages/2018

Kaufman, E., & Freitekh, S. (2021). Transforming a dream into reality: Palestinian and Israeli youth struggling across the divide for universal human rights, democracy, and peace. PIJ.ORG. https://www.pij.org/articles/2098/transforming-a-dream-into-reality-palestinian-and-israeli-youth-struggling-across-thedivide-for-universal-human-rights-democracy-and-peace

Kelman, H. C. (2018). *Transforming the Israeli-Palestinian conflict: From mutual negation to reconciliation* (P. Matter & N. Caplan, Eds.). Routledge.

Klein, J. P., Goertz, G., & Diehl, P. F. (2006). The new rivalry dataset: Procedures and patterns. *Journal of Peace Research, 43*(3), 331–348. https://doi.org/10.1177/0022343306063935

Leshem, O. A. (2017). What you wish for is not what you expect: Measuring hope for peace during intractable conflicts. *International Journal of Intercultural Relations, 60*, 60–66. https://doi.org/10.1016/j.ijintrel.2017.06.005

Leshem, O. A. (2019). The pivotal role of the enemy in inducing hope for peace. *Political Studies, 67*(3), 693–711. https://doi.org/10.1177/0032321718797920

Leshem, O. A., & Halperin, E. (2020a). Hope during conflict. In S. C. van den Heuvel (Ed.), *Historical and multidisciplinary perspectives on hope* (pp. 179–196). Springer International. https://doi.org/10.1007/978-3-030-46489-9_10

Leshem, O. A., & Halperin, E. (2020b). Lay theories of peace and their influence on policy preference during violent conflict. *Proceedings of the National Academy of Sciences*. https://doi.org/10.1073/pnas.2005928117

Leshem, O. A., & Halperin, E. (2020c). Hoping for peace during protracted conflict: Citizens' hope is based on inaccurate appraisals of their adversary's hope for peace. *Journal of Conflict Resolution, 64*(7–8), 1390–1417. https://doi.org/10.1177/0022002719896406

Leshem, O. A., & Halperin, E. (2021). Threatened by the worst but hoping for the best: Unraveling the relationship between threat, hope, and public opinion during conflict. *Political Behavior, 20*.https://doi.org/10.1007/s11109-021-09729-3

Lupovici, A. (2013). Pacification: Toward a theory of the social construction of peace. *International Studies Review, 15*(2), 204–228. https://doi.org/10.1111/misr.12032

Malley-Morrison, K., Mercurio, A., & Twose, G. (2013). *International handbook of peace and reconciliation*. Springer.

Manekin, D. (2017). The limits of socialization and the underproduction of military violence: Evidence from the IDF. *Journal of Peace Research, 54*(5), 606–619. https://doi.org/10.1177/0022343317713558

Maoz, I., & McCauley, C. (2008). Threat, dehumanization, and support for retaliatory aggressive policies in asymmetric conflict. *Journal of Conflict Resolution, 52*(1), 93–116. https://doi.org/10.1177/0022002707308597

Morris, B. (2009). *One state, two states: Resolving the Israel/Palestine conflict*. Yale University Press.

Navot, D., Rubin, A., & Ghanem, A. (2017). The 2015 Israeli general election: The triumph of Jewish skepticism, the emergence of Arab faith. *Middle East Journal, 71*(2), 248–268. https://doi.org/10.3751/71.2.14

Nijs, T., Stark, T. H., & Verkuyten, M. (2019). Negative intergroup contact and radical right-wing voting: The moderating roles of personal and collective self-efficacy. *Political Psychology, 40*(5), 1057–1073. https://doi.org/10.1111/pops.12577

Oakeshott, M. (1996). *The politics of faith and the politics of scepticism*. Yale University Press.
O'Malley, P. (2016). *Two-state delusion: Israel and Palestine: A tale of two narratives*. Penguin Publishing Group.
Palestinian Central Bureau of Statistics. (2017). Main statistical indicators in the West Bank and Gaza Strip, 2017 Census. http://www.pcbs.gov.ps/Portals/_Rainbow/StatInd/StatisticalMainIndicators_E.htm
Rouhana, N. N. (2004). Group identity and power asymmetry in reconciliation processes: The Israeli-Palestinian case. *Peace and Conflict: Journal of Peace Psychology*, *10*(1), 33–52. https://doi.org/10.1207/s15327949pac1001_3
Rouhana, N. N., & Bar-Tal, D. (1998). Psychological dynamics of intractable ethnonational conflicts: The Israeli–Palestinian case. *American Psychologist*, *53*(7), 761.
Sagy, S., & Adwan, S. (2006). Hope in times of threat: The case of Israeli and Palestinian Youth. *American Journal of Orthopsychiatry*, *76*(1), 128–133. https://doi.org/10.1037/0002-9432.76.1.128
Sethi, S., & Seligman, M. (1993). Optimism and fundamentalism. *Psychological Science (Wiley-Blackwell)*, *4*(4), 256–259.
Shani, M., & Boehnke, K. (2017). The effect of Jewish–Palestinian mixed-model encounters on readiness for contact and policy support. *Peace and Conflict: Journal of Peace Psychology*, *23*(3), 219–227. https://doi.org/10.1037/pac0000220
Shelef, N., & Zeira, Y. (2017). Recognition matters!: UN state status and attitudes toward territorial compromise. *Journal of Conflict Resolution*, *61*(3), 537–563. https://doi.org/10.1177/0022002715595865
Shulman, D., Halperin, E., Elron, Z., & Reifen Tagar, M. (2021). Moral elevation increases support for humanitarian policies, but not political concessions, in intractable conflict. *Journal of Experimental Social Psychology*, *94*, 104113. https://doi.org/10.1016/j.jesp.2021.104113
Sulitzeanu-Kenan, R., & Halperin, E. (2013). Making a difference: Political efficacy and policy preference construction. *British Journal of Political Science*, *43*(02), 295–322. https://doi.org/10.1017/S0007123412000324
Thiessen, C., & Darweish, M. (2018). Conflict resolution and asymmetric conflict: The contradictions of planned contact interventions in Israel and Palestine. *International Journal of Intercultural Relations*, *66*, 73–84. https://doi.org/10.1016/j.ijintrel.2018.06.006
Tilley, V. (2010). *The one-state solution: A breakthrough for peace in the Israeli-Palestinian deadlock*. University of Michigan Press/Eurospan [distributor].

5

Hope and Activism

Oded Adomi Leshem, Shanny Talmor, and Eran Halperin

Those whose hope is weak settle for comfort or violence.

—Erich Fromm (1968)

The year 2005 presented a surprising change in Israel's policy toward the territories it conquered in 1967. After more than thirty-five years of military control, Israel's government decided to pull out from the Gaza Strip.[1] This large-scale operation, dubbed the Disengagement, entailed, on top of a complete military withdrawal, the eviction of 8,000 Jewish Israeli settlers, some of them living in the Strip for almost three decades. In the months before the disengagement, tens of thousands of Israelis took to the streets, some adamantly supporting the disengagement and some fiercely opposing it. During August and July 2005, the public debate over the disengagement was so tense that some felt Israel was on the verge of a civil war (Rynhold & Waxman, 2008).

Many Jewish Israelis living in the vicinity of the Gaza Strip believed that the disengagement would bring peace and quiet to Gaza and the surrounding Israeli villages. This prognosis proved wrong. The hostilities in and around Gaza persisted after the disengagement, with more than 5,000 Gazans and 100 Israeli dead since the withdrawal in August 2005.[2] Moreover, the "Gaza Issue" seemed even more intractable after the disengagement. The split between the Hamas-ruled Gaza Strip and the Fatah-ruled West Bank, as well as the different policies Israel implements in each region, created new obstacles that seem insurmountable as the years pass.

[1] Four Israeli settlements in the North of the West Bank were also dismantled as part of the disengagement.

[2] https://www.ochaopt.org/data/casualties. Several factors can explain why clashes continued after the disengagement. The two apparent ones are the rise of the Hamas in Gaza and Israel's strict blockade of the Strip. However, other reasons can be provided, such as Palestinians' belief that withdrawing from Gaza helped Israel increase its grip in the West Bank.

Hope Amidst Conflict. Oded Adomi Leshem, Shanny Talmor, and Eran Halperin Oxford University Press.
 DOI: 10.1093/oso/9780197685303.003.0005

Despite the unfavorable conditions, a few Palestinians and Israelis established a small cross-border grassroots nongovernmental organization (NGO) called Other Voice. The NGO includes Palestinian Gazans and Israelis from the surrounding area who together call for an end to hostilities. During peaks of confrontation, Palestinian and Israeli members of the NGO stay in touch and support one another. Other Voice members report that during these challenging times, when the Strip is bombarded by Israeli planes and Kassam missiles are flying toward Israeli cities, the solidarity between the activists provides a sense of hope. However, this hope is challenged when members voice their agenda publicly. Israeli activists from Other Voice standing with "End the War" signs in street junctions often get harassed and sometimes violently attacked by angry onlookers. Palestinian members of the NGO are frequently interrogated by Hamas officials for keeping in touch with Israelis.

At least intuitively, it seems reasonable to assume that the level of hope for peace among peace activists from Other Voice and other NGOs is relatively high.[3] Otherwise, why would these activists dedicate their lives to peace and conciliation? More specifically, it could be easily argued that, compared to non-activists, activists' wishes and expectations for peace are likely to be high, consequently driving their commitment and hard work. However, in the face of ongoing conflict and endless hostilities, peace activists are probably those most susceptible to despair. Tirelessly working for peace without seeing any progress is likely to have a detrimental impact on hope.

Furthermore, peace activists are considered by many ingroup members as foolish idealists or even traitors who think about the enemy's interests before the interests of their group (Shulman, 2007). The high social price activists pay adds a considerable challenge to activists' hope for peace. Looking at it this way, it could be the case that peace activists working in intractable conflict are, in fact, those most hopeless.

Are peace activists hopeful or hopeless? How can we reconcile these two opposing possibilities? This chapter, written with Shanny Talmor and Eran Halperin, explores this question by asking the activists themselves. Shanny conducted in-depth interviews with Palestinian and Israeli peace activists from four peace NGOs to understand how hope functions at the grassroots level. The basic rationale for this unique inquiry is that hope can be understood best when challenged the most. As we unfold in this chapter, the hope of

[3] Consistent with the bidimensional model, I use the word "hope" to refer simultaneously to the two dimensions. However, when I want to refer to only one dimension I do not use "hope" but the specific name of the dimension: "wishes" or "expectations."

Palestinian and Israeli peace activists is constantly tested. We wanted to hear from them, male and female activists working in Gaza, the West Bank, and Israel, if and how hope makes them tick.

Two main interests guided the design and analyses of our research. The first was to understand hope at its points of extremes. Because peace activists are heavily invested in advancing peace, their hopes can reach great peaks when peace seems near and extreme lows when it seems furthest away. As we have learned from our interviewees, sometimes the fluctuation in hope happens fast. For example, after their hopes had spiked during a successful cross-border meeting, activists' hopes often plummeted upon their return to their communities that showed no desire for peace or any belief in its feasibility. Our second goal was to provide in-depth insights into our quantitative research on hope amidst conflict. Qualitative analyses complement quantitative approaches by revealing nuances and meaning-making processes that numerical measures have difficulties detecting (Mills & Birks, 2014). Indeed, qualitative research on hope has an impressive legacy (e.g., Averill & Sundararajan, 2005; Breznitz, 1986). However, qualitative work on hope in the context of conflict is scant (for exceptions, see Dowty, 2006; Jamal & Lavie, 2021).

Our qualitative path proved rewarding for deepening our understanding of hope and advancing research on activism in intergroup conflict. The interviews with Israeli and Palestinian activists provided qualitative evidence supporting the bidimensional model while also questioning some assumptions about hope as a motivator for action. Furthermore, exploring hope among peace activists who exhibit admirable persistence in the face of a seemingly hopeless reality revealed the boundaries of hope, even in the eyes of the activists themselves. We also present what might be considered a counterintuitive finding. At least to some extent, hope seems to be one of the end goals of political activism, not only political change itself.

This chapter begins by introducing the topic of activism and its link with hope and other psychological motivators. We then briefly present the tradition of peace activism in the context of the conflict in Israel-Palestine and expand on the four peace NGOs investigated in this study. Next, we provide details on the methods employed in the interviews and the analysis strategy of their content. We then turn to our main findings on the role of hope in grassroots peacebuilding, followed by a discussion on their implication for theory and practice. The chapter ends by linking knowledge from this study to the book's broader premises. We note that the interviews provided many insights about hope and activism, some of which are not presented here due to length limitations.

Activism and Hope

Political activism is carried out by groups of people coordinating toward attaining social and political goals (Nardini et al., 2021). Several models in political psychology explain why people become active in social and political movements (e.g., Thomas et al., 2012; van Zomeren, 2013). Among the central factors that motivate people to take action is a sense of injustice, anger, strong identification with the group, and firm beliefs in group efficacy. Hope has also been studied as a motivator of participation in activism (Cohen-Chen & Van Zomeren, 2018; Courville & Piper, 2004; Greenaway et al., 2016; Leshem, 2019; Wlodarczyk et al., 2017) but mainly as a mediator between group efficacy and participation in collective action. Moreover, studies that explored the link between hope and activism did not utilize the bidimensional approach. Therefore, it is difficult to identify the primary driver of collective action: Is it the wish for social change or the expectation that social change is possible?

Another concern regarding hope's role as a motivator for activism comes from studies on environmental activism that show that hope has no direct influence on collective action (Furlong & Vignoles, 2021; van Zomeren et al., 2019). Furlong and Vignoles (2021) suggest that hope does not predict participation in environmental activism because hope has a *palliative* function and is not necessarily a *problem-focused* construct. Stated otherwise, people might experience hope only to ease their anxieties rather than change the social or political conditions that create these anxieties. This argument is in line with Breznitz's observation (1986) that hope can be an end goal in and of itself, not simply a path to achieve the hoped-for goal. Indeed, hope is an effective coping mechanism (Lazarus, 1999) and has inherent psychological advantages (Burger et al., 2020; Litt et al., 1992). It follows that, in some situations, achieving the political goal can become a secondary aim, while attaining hope becomes the primary target. Another explanation about why researchers found no association between hope and environmental activism is presented in Chapter 7.

The preceding review suggests that the link between hope and activism requires additional scrutiny. Several possible relations between hope and activism can be postulated. The first, though unlikely, is that hope has nothing to do with peace activism and that other factors (e.g., anger at the system, moral responsibility) are the motivators of peace activism. Another option is that activism has a nuanced relationship with hope that, so far, has not been described. Perhaps the bidimensional model of hope might assist. For instance, looking separately at each dimension can reveal if peace activism is more about the desire to end the conflict or the expectations that the conflict can end. Still another relationship can be suggested: that peace activism is intended, first and foremost, to bring hope, not peace. If this is the case, the connection between hope and activism is primarily a reversed one, where the involvement in peace activism is directed to safeguard an existential need, the need to hope. These postulations are provocative, perhaps counterintuitive. Nevertheless, we believed that we could provide some answers to these questions by talking with activists working for peace in one of the most intractable conflicts in the world.

Peace Activism in Israel-Palestine

From American activists protesting against their country's involvement in Vietnam to the anti-proliferation protesters in Europe, peace activism played an essential role in the trajectories of war and peace processes (Schwebel, 2008; Small, 1988). The history of peace activism is rife with diverse techniques, initiatives, and activities (Perry, 2011; Small, 1988). Activists have marched, coordinated sit-ins, and advocated for dissent and civil disobedience in times of war and conflict. Not surprisingly, studies have shown that peace activists' values and moral convictions are solid and salient (Schwebel, 2008). Their cherished values, so central to their life, provide a sense of strength but also considerable vulnerability. Strengths come in the form of their choices, often at the risk of social disapproval. However, their strong moral convictions are also a source of vulnerability. When their cherished values do not materialize, activists are the ones who will be discouraged the most.

Some details about the history of the peace movement in Israel-Palestine are helpful at this stage. First, within Israel, there have been some organizations and movements urging Israel to adopt a dovish stand toward neighboring Arab counties and the Palestinians. Specifically, since 1967, the Israeli peace movement has sought to convince the Jewish Israeli public and leadership that Israel should relinquish conquered land in return for peace (Hermann, 2009). One of the most notable movements is Peace Now, which organized in 1982 what was then the largest rally in Israeli history. The rally protested Israel's involvement in Lebanon in the wake of the Sabra and Shatila Massacres and called for an end to the war. Other organizations, like the Four Mothers, who were instrumental in pushing Israel to withdraw from South Lebanon, and the track-two diplomatic effort Geneva Initiative, have also influenced public discourse in Israel, at least in certain areas.

In addition to peace organizations, several Israeli human rights NGOs have been working against Israel's discriminatory policies and oppression of Palestinians in the Occupied Palestinian Territory (OPT). Betzelem, Hamoked Lahaganat Haprat, and Yesh Din are some human rights NGOs trying to provide legal aid to Palestinians and expose Israel's human rights violations. However, most importantly, peace movements and humanitarian NGOs are negligible in number and impact. The number of active members is low, budgets are small, and their work is hardly mentioned in the media. Public sentiment is generally hostile toward peace and human rights movements in Israel, with some Israelis calling to ban these organizations altogether (for an overview, see Hermann, 2009).

In the OPT, the picture is very different. First, peace movements do not exist in the traditional form within Palestinian society because the entire populace is absorbed in a struggle for self-determination. However, among the general widespread support for violent resistance, some Palestinian groups also advocate for a nonviolent struggle, including collaboration with Israelis who oppose the occupation. Taayush, Taghyeer, and Combatants for Peace are some examples.

Binational Peace NGOs in Israel-Palestine

A unique form of peace activism is the *binational model*, which includes the collaboration of activists from the two sides in a united effort to stop the violence and promote reconciliation, justice, and peace. Among the few peace NGOs working in Israel-Palestine, only a handful can be considered binational. Needless to say, conducting routine activities is exceptionally challenging for binational NGOs. The logistics are almost impossible. Palestinians rarely get permits to enter Israel proper, and Jewish Israelis cannot enter Palestinian-populated areas in the OPT. Israelis and Palestinians can only meet in area B in the West Bank, so joint activities are often conducted there.[4]

Moreover, many Palestinians are reluctant to work with Israelis for fear that collaborating with Israelis will normalize the abnormal reality of occupation and oppression. In addition, many Israelis are afraid to work with Palestinians, who are portrayed in the Israeli media as terrorists. These challenges are heightened during times of escalation, when violence is on the rise and the power disparities between the groups become even more noticeable. In times of calamities (and unfortunately, these are frequent), the mere integrity of binational groups may be severely threatened. Due to these and

[4] B area consists of about 22% of the West Bank. In B area, civilian matters are managed by the Palestinian Authority while security matters are the responsibilities of Israel.

other challenges, binational peace NGOs are rare. However, because of their binational structure and the extreme challenges they face, binational peace NGOs are a perfect fit for the study of hope and hopelessness. Hope is constantly challenged in binational peace organizations. On the one hand, successful programs run by binational organizations create small-scale peace realities likely to boost hope among members. On the other hand, because peace is getting farther despite their great efforts, members of these NGOs are likely to be despaired.

The Current Study

We focused on four NGOs for our qualitative investigation of hope and peace activism. These grassroots NGOs, all working on issues at the heart of the dispute, were chosen based on several criteria. The first is their grassroots nature, which means that most members are volunteers. Volunteering in these organizations requires perseverance, resourcefulness, and, in some situations, courage. We thus believed that the grassroots nature of the NGO would heighten the role of hope in members' activism. The second criterion is their binational structure, which enabled us to examine hope in a highly challenging environment and interview Israeli and Palestinian activists who work together as equals. Focusing on binational NGOs offers a methodological advantage because it improves comparability across societies and contexts. That said, it should be noted that, due to the extreme power disparities, the conditions in which Israeli and Palestinian activists work are not equal at all. Eventually, asymmetry affects the flexibility, scope, reach, and impact of activism in Israel and the OPT. The last criterion was the NGOs' geographical spread. The four NGOs work in the entire area between the River and the Sea, including the West Bank, Gaza Strip, East Jerusalem, and Israel proper. Two NGOs have received some media attention, and the two others are relatively unknown. This enabled us to explore different perspectives on activism and hope. Following is a short description of the organizations.

Combatants for Peace was established in 2006 by Israeli and Palestinian excombatants who took an active part in the armed violence (as soldiers in the Israeli Defense Force [IDF] or as members of Palestinian militias) and decided to throw their weapons down in a joint call for peace and justice. The NGO consists of several dozen members who advocate for an end to Israel's occupation and a just and viable peace between the peoples. While writing this chapter, members of Combatants for Peace were attacked by Israeli soldiers while the activists delivered drinking water to isolated Palestinian villagers in the West Bank. In 2017 and 2018, Combatants for Peace were among the nominees for the Nobel Peace Prize.

The Parents Circle (formally called the Israeli-Palestinian Family Forum of the Bereaved) was founded in 1995 by Palestinians and Israelis who lost close family members in the conflict. This NGO wishes to influence public discourse toward support for peace and tolerance through dialogue groups, media campaigns, and other grassroots activities. They seek to use their unique position as bereaved people to break stereotypes and foster cross-border understanding. Their annual Alternative Memorial Ceremony commemorating all those who died in the conflict is attended by thousands.

Shorashim-Judur is a small and local network of Palestinian and Jewish Israeli settlers living in the Gush-Etzion region in the OPT. The NGO's mission is to promote trust, dialogue, nonviolence, and transformation between the residents of the locality and spread the message of co-existence to others in Israel-Palestine. Members of the NGO believe in the rootedness of Palestinians and Jews in the Holy Land and the equitable distribution of resources.

Finally, *Other Voice* is a local organization connecting Palestinians from Gaza and Israelis from the area surrounding the Gaza Strip. As residents of one of the most volatile regions in the world, members of the group seek to raise public awareness of the price of the conflict and the need to reach a sustainable solution to the dispute. The group maintains a supportive dialogue between members from both sides of the border and tries to distribute a joint call to end mutual violence.

Research Methods

Semi-structured in-depth interviews were conducted with ten Jewish Israelis and ten Palestinians from the four NGOs (eleven males and nine females). The interviews were conducted between March 2020 and February 2021, took about an hour and a half each, and were done in Hebrew or English, depending on the interviewee's request. To facilitate the exploration of our main research questions, the interviews were comprised of two parts. In the first part, we did not reveal that hope was the focus of our research. This allowed participants to speak freely about their activism without us prompting the subject of hope and enabled us to detect whether hope was brought up spontaneously by the interviewees and, if so, in what contexts and capacities. We were also attentive to the use of hope as "wish" or "expectations" and how the dimensions were linked to activism. We then moved to explicitly ask participants about hope, activism, and their connection. Here we were interested to know how interviewees defined the relationship between their activities and hopes. At some point, we also mentioned the distinction between wishes and expectations and asked them to elaborate on the connection of each dimension to their activism. From this stage, we wanted to learn about the centrality of hope in their lives and activities, especially in times of escalation.

The interviews were analyzed using thematic analyses (Braun & Clarke, 2006). This method helped us arrange the entire body of the interviews (more than 300 pages of transcribed interviews) based on the central themes emerging in the text, thereby creating a detailed yet organized overview of the corpus. More specifically, after transcribing the recordings, we coded all interviews for explicit and implicit references to our main topics of interest (e.g., hope, despair, wish, expectations, activism). We then consolidated these references across interviews and arranged them according to themes and our main research questions. Next, we reduced the number of themes to those that best answer our questions and are salient enough across interviews. The results presented below are based on analyzing and interpreting these central themes. To ensure anonymity, we changed the names of all participants. However, their age, gender, and nationality (in parenthesis) have not been changed.

Results

Spontaneous Use of Hope

In most cases, hope emerged in the interviews without our prompt. Interviewees mentioned hope and its connection to activism spontaneously, which helped us investigate its role in participants' work without directly asking them about it. Here, we outline two salient and interesting ways interviewees talk about hope. The first is *hope as a basic need*, where interviewees mentioned hope as an essential component in their lives and, more generally, in people's lives. The second theme that emerged relates to the dimensions of hope. We noticed that interviewees sometimes used hope to express their wishes for peace and, at other times, their expectations that peace would materialize. Both dimensions seemed relevant to activists' notion of hope and its relationship with their activities.

Hope as a basic human need. As elaborated in previous chapters, looking at hope as a basic human need was proposed by thinkers from Immanuel Kant to Victor Frankl. It seems that our Palestinian and Israeli participants also talked about hope as an essential component in their lives and the lives of humans more generally. The overall theme of hope as a basic need was manifested explicitly and implicitly. Some interviewees explicitly mentioned hope's centrality to human life. For example, Jaber (48, Male, PAL) stated his belief that, without hope, one is left lifeless (1). He also claimed that hope is one of the factors distinguishing humans from animals. Thus, without hope, one turns into lower animals that have no aspirations or deeper meaning in life. Looking at the centrality of hope from a different angle, Amir (45, Male, ISR) defined hope as an essential compass for his thoughts and actions, something he is unwilling to give up (2).

(1) "We have to keep moving on. A human being without hope is a human being without life. You become like an animal: eat, sleep, and nothing more."

(2) "I am someone that clings to hope. I am not willing to forfeit hope . . . it is something I count on, depend on . . . on hope. . . . Many people ask me if I'm optimistic or pessimistic. I try not to think in those terms, but I am not willing to lose hope. This means that the word hope is a compass for me."

Notice that apart from asserting that hope is a valuable compass in his life, Amir (2) said he tries not to think in terms of optimism and pessimism. There were other examples where interviewees agreed that hope might be essential for human survival but do not think and act according to this assumption. The words of Roi (72, Male, ISR) are a perfect example of this duality (3).

(3) "I don't think anyone can live without hope. But I don't act by thinking in terms of hope. I act based on what needs to be done. I ask myself: How do I give hope to others?"

Throughout many interviews, hope was mentioned as an essential component in participants' lives. However, some talked about the centrality of hope more implicitly. For example, Gabriela (68, Female, ISR) implied that hope is central to her life by demonstrating her perseverance and rejection of hopelessness (4).

(4) "This is how I work: I have a plan A, if it doesn't work, I turn to plan B, and if that doesn't work, I turn to plan C [laughs] . . . Not everything I wish for materializes, but if it doesn't, I say OK, let's try it another way, and so on and so on. Maybe not now, maybe in a month, maybe it's not good timing . . . I don't get discouraged."

Gabriela described her activism as an exercise of persistence and ended with rejecting discouragement as an option. Though implicit in her words, the centrality of hope in her activities was apparent. The above examples suggest that activists, like prominent thinkers and philosophers of hope (e.g., Fromm, 1968; Hume, 1740/2003; Tillich, 1965), view hope as a fundamental aspect of human existence. However, a more nuanced look is in place, especially regarding how the two dimensions of hope operate in peace activism.

Hope as a bidimensional construct. By now, the reader should be familiar with the idea that in everyday speech, hope sometimes signifies one's desires and sometimes one's expectations of fulfillment. The question put forth in this section is, what can we learn about the dynamics of hope in activists' lives, given hope's different meanings? To answer this question, we analyzed the text by focusing on the meaning or meanings of the word hope. We were first interested in determining whether the interviewee used hope to refer to his or her wishes for peace or to his or her expectations that peace will materialize. We then examined how each dimension functioned in the larger context of participants' life and activism.

As expected, we found that both dimensions were represented in the interviews. However, the roles and dynamics of hope were very different when hope was used to represent activists' wishes versus their expectations for peace. First, it seems that when hope signified activists' "wishes" for peace, it was high, stable, and unquestionable. High wishes for peace were often mentioned as a critical motivator for joining activities. For instance, when asked about her thoughts before joining the peace NGO, Lina (43, Female, PAL) said,

(5) "Yeah. I was thinking of peace. . . . I didn't know how to be part of it, but I hoped [for peace] because I didn't want to lose any of my children or anyone from my family members. I wanted peace for all of the people because we felt tired. We feel that we deserve a better life for us, for our children and grandchildren."

Here Lana "hopes" that peace will prevail for the safety and well-being of her family. Notice that Lana used hope to express her wish for peace, not necessarily her assessment of its likelihood. Further note that Lana used "want" as a synonym to hope, suggesting that Lana was talking about her wishes, not her expectations for peace. Hope as "wish" is also used to describe future plans and dreams of change. Miriam (40, Female, PAL), who works with Gazan youths, described several peace projects she wants to implement in the Gaza Strip. She enthusiastically talked about the programming and her "hopes" for the future.

(6) "That's what I really want to do more. My hope is to create or to be able to have all the financial resources to create, to develop a program to help the youth peacebuilding programs."

Looking at the entire corpus, it seems that, almost without exception, when "hope" was used to signify interviewees' wishes for peace, it was described as high and enduring. This hope (i.e., wish) began before the interviewee joined the peace NGO and extends into the future, durable, firm, and decisive.

However, a very different dynamic emerged when activists used the word "hope" to refer to their expectations for peace. When hope represented their expectations, hope was indecisive and often low. For example, several participants talked about how their hope for peace (i.e., expectations for peace) oscillates depending on political circumstances. During certain times, when there are small signs of advancement, their hope (i.e., expectations) for peace rises.[5] Yet when these advancements reached a dead end, their hopes dropped. For instance, Nur (25, Male, PAL) told the story of why his parents moved from Jordan back to Palestine in the 1990s (7).

(7) "So my family decided to move back during the Oslo years . . . after the Oslo Peace was signed. It was a sign of hope, and a sign for building a new future so they can come back from Amman, lived here in Bethlehem, and started, you know, a new future."

Nur then revealed that these hopes (i.e., expectations) were soon shattered in the wake of the Second Intifada. The Palestinian uprising in the Second Intifada was met with Israel's strong military response that devastated the lives of many Palestinian families, including Nur's. It seems that hope as expectations is fragile and frail compared to the resoluteness of hope as wishes. Hope's fragility is well described by Bassam (24, Male, PAL). When Bassam was asked if he saw himself as a hopeful person, he replied,

(8) "Not necessarily. No, not now, certainly, I mean, not always. I have hopes, yeah, but I'm not always hopeful because one of the things that you learn in life is that when you have expectations and hope for something, and you don't get that something, even in your personal life, not just, you will feel worse than before."

Bassam's description of his hope demonstrates the distinction between hope as wishes and hope as expectations. Bassam says he has hopes (i.e., wishes) for peace, but he is not necessarily hopeful (i.e., expects peace to materialize). Many activists offered similar testimonies; their wish for peace is evident, but their expectations are not. Bassam provided a compelling reason

[5] The mid-1990s are often mentioned by the more seasoned activists as a time of great hope, the Second Intifada that began in October 2000 as a time of extreme hopelessness.

to be cautious about having high expectations for peace, namely, the need to protect oneself from dashed hopes. In several places in the interviews, activists criticized people who were sure that peace would come. The interviewees claimed that high expectations for peace reflect naiveté and detachment from political reality.

Overall, activists' expectations for peace seem quite shaky and depend on external cues, like the escalation or de-escalation of conflict. Are there events or processes that influence activists' expectations for peace for the positive? Specifically, we were interested in finding evidence in the texts about whether the activities increased participants' belief in the possibility of peace. Looking at the corpus of interviews, it seems that participants often claim that their hopes (as expectations) increase after encounters with fellow activists from the other side of the border. Gefen (50, Female, ISR), for example, shared her feelings:

(9) "Every small interaction with the other side brings a feeling of elevation. Wow! How amazing are those people! Planning activities together, working with them in the field. . . . Sure this brings hope! How can't it make you more hopeful?"

Here Gefen revealed how her expectations for peace increased after a binational activity. Seeing with her own eyes a fruitful contact between the two peoples increased her belief that peace is possible and thus raised her hopes. Positive peacebuilding experiences, especially successful experiences from binational activities, seem to boost activists' expectations for peace. During these activities, peace seems possible, perhaps inevitable, if only for a fleeting moment. Conversely, interviewees also mentioned that unsuccessful encounters impacted their expectations for peace in the opposite direction. Each time the binational activity failed, their hopes (as expectations) received a blow. Before looking more closely at the link between hope and activism, it will be helpful to look at Salem's (48, Male, PAL) comment about the tension between activists' high wish for peace and, sometimes, their low expectations that peace will come about (10).

(10) "Ok, you keep one leg in reality and one leg in the dream, and that's where I am moving between legs. One leg in the reality that is grounded and one in the dream, yeah, or the vision"

In his own words, Salem seemed to refer to the two components of hope. The first is the "dream," which relates to the wishes and aspirations for peace, and the second is the "reality on the ground," which relates to the assessment of the feasibility of peace. The two legs are essential, as Salem asserted. Activists cannot live the dream of peace in isolation from harsh reality. They must acknowledge that the reality is tough, even if it means that one's expectations for peace must be kept low. Naïveté is not helpful. However, the wish for peace, the "dream," is equally essential. Without it, no one will invest time or energy in peace.

Overall, it appears that activists have very high wishes for peace, which are, according to their statements, uncontested and unshakeable. High wishes for peace correspond with data from the general Palestinian and Jewish Israeli population (Leshem, 2017; Leshem & Halperin, 2020b; see also Chapter 4), although activists' wishes for peace are probably on the extremely high end of the scale. Activists' expectations for peace, on the other hand, are described by the interviewees as feeble and indecisive. Some also shared that they are pessimistic about the chances of peace in Israel-Palestine but that their commitment to peace is independent of this pessimism. In fact, when interviewees used hope as "expectation," some dismissed the notion that their actions were based on their hopes. It is a mistake, in their eyes, to base activism on expectations for peace as these expectations depend on external factors they cannot control. Indeed, if activism were based on expectations for peace, it would completely cease during escalation and heightened hostilities when it is most needed.

Two interim insights can be drawn about activists' expectations for peace. The first is that activists' beliefs in the possibility of peace are not exceptionally firm, at least as reported by the Israeli and Palestinian activists in our study. The second point is that the expectation dimension of hope should not be the primary motivator for activities due to its fragility.

The Links Between Hope and Activism

Most interviewees described the relationships between hope (in either dimension) and activism as reciprocal. On the one hand, participants talked about hope as a catalyst for activism. Adi (72, Female, ISR) (11) and Jaber (48, Male, PAL) (12), for example, explicitly mentioned hope as the cause of activism.

(11) "It's only due to hope that you can work for the cause. It gives you the strengths, it gives you the strength in places when you feel that it's not going so well. But then you have hope, so you persist."

(12) "Hope is what drives a person who believes in peace. Hope is the fuel, your engine. Without it, you lose everything."

On the other hand, activism was often described as *the cause* of their hope. As Gefen's remark above (9) exemplifies, the activities themselves stimulate activists' hopes. Another example is Nadim's (24, Male, PAL) account of the connection between activism and hope.

(13) "Inside the Israeli society and the Palestinian society, the gap is not getting smaller . . . I feel it's getting wider . . . but what makes me hopeful is that activists from both sides meet. Even though there are small numbers maybe. But there are people who are realizing in the middle of this madness that we must do something, that we have to stop this."

Both causal directions are evident in the interviews, indicating a reciprocal relationship between hope and activism. However, looking at the text more nuancedly, we also detected "interruptions" in the perceived reciprocity between hope and activism. By interruptions we mean places where the mutual, reciprocal connection between hope and activism is broken. As stated, some people explicitly said that their hopes do not necessarily lead to action. In other interviews, participants mentioned other factors that propelled their activities. For example, Amjad (50, Male, PAL) and Limor (67, Female, ISR) claimed that their activism could not be explained by hope but by the values they hold and the understanding that they have a responsibility to act on these values. Limor talked about responsibility to the next generations (14), and Amjad talked about moral obligation (15).

(14) "What guides me is that one day my grandkids or my great-grandkids will ask me, 'Grandma, where were you when all this happened? What did you do?' And I will have to give them answers."

(15) *Q*: Do you think that hope is also a motivator of your activism? *Amjed*: "No, my motivation is my obligation. If I needed to be motivated by hope, I would not get up from my bed. I get up from bed due to feelings of moral obligation to do what I do. Sometimes I do it with less hope, and sometimes I do it with more hope. But I do it because I must."

In these examples, hope is not mentioned as the key driver of action. In Amjad's case, hope was explicitly excluded from the list of motivators of activism. A more nuanced reading of the text suggests that Amjad referred to the expectation component of hope. Amjad may have meant that if the possibility of reaching a peace agreement were his primary motivator for change, he would not have got out of bed. Even more interestingly, Amjad claimed that discouragement and despair, the opposites of hope, motivate his activism (16).

(16) "When I'm despaired, it actually motivates me to act. And then acting makes me feel a little less despair. I meet people [other activists], and I see that there is some development. The change is very incremental, but at least it's a change in the right direction."

Amjad's statement completely breaks the reciprocity between activism and hope. He goes to the meetings to receive encouragement and replenish his hopes, but, as he and others testified, hope does not necessarily motivate his activism. Some activists mentioned other motivators, like responsibility, while Amjad mentioned despair. Roi (71, Male, ISR) expressed a similar rejection of hope as a driver of his activities, citing commitment as the source of his actions.

So, activism leads to hope, but hope does not necessarily lead to activism. How could this be explained? Based on the notion that hope is a basic human need, one possible explanation is that for Amjad and perhaps others who spoke in a similar vein, maintaining a sense of hope is the primary goal of activism. As stated by many participants, activism replenishes their hopes. When members meet colleagues from across the border, their belief in the possibility of peace increases. The increased expectations feed into their hope, which must be replenished in the face of ongoing adversities. In other words, regardless of whether activism can bring peace, activism brings hope.

Discussion

Why do activists do what they do despite the significant challenges and the hefty price they often pay? This question is particularly relevant for peace activism in a century-old conflict. It appears almost irrational for individuals to devote substantial amounts of time and energy to pursue peace in a situation deemed irresolvable. Furthermore, peace activists often risk their reputations, day jobs, and personal security by associating themselves with their activities. Why would they do it?

It certainly stands to reason that one of the drivers of their choice to engage in activism is hope. Outside observers intuitively deduce that one of the unique factors of activists is that they think peace is possible, perhaps inevitable, and so pursuing peace makes sense, at least in their own eyes. However, also intuitive is that, in working to transform a century-old conflict, peace activists are likely to experience deeper feelings of despair compared to non-activists. At times of escalation and calamity, and when targeted as traitors by kin and ingroup members, activists' hopelessness is profound, possibly more so than those of non-activists who make no effort to end the conflict.

Interviews with twenty Israelis and Palestinians working on the frontline of peace activism shed light on this and other puzzles concerning the role of hope in collective action. First, although not tested quantitatively, we have found no evidence that Israeli and Palestinian activists have exceptionally high expectations for peace. From the interviews, we see that activists' belief in the feasibility of peace is unstable and low and possibly no different from the expectations for peace in the general society. Indeed, why should activists have higher expectations for peace? In fact, given their relentless work for peace in the face of a never-ending conflict, their feelings of despair should be more profound than those of non-activists. Moreover, several interviewees explicitly rejected the notion that their activism is driven by their expectations for peace, stating that they would not work for peace if this were the case. Throughout the interviews, the typical response to the question about the chances of peace was that it depended on external factors currently pointing in the direction of further escalation. Participants admitted that their expectations could not provide a solid motivational backbone for their peacebuilding work.

Is there no connection between hope and activism? From the interviews, it seems there is a strong connection between activism and hope, but with the wish rather than the expectation dimension. In most interviews, activists' unequivocal wishes for peace stood out as a central driver of their choices. Both explicitly stated and implicit in the subtext, participants' desires for peace appear sound and robust. This stable and unyielding wish for peace, coupled with a sense of responsibility and commitment to their society and offspring, seems to drive their activism. In short, among our interviewees, peace activism was driven more by their aspiration for peace than by the belief in its feasibility.

The second point to be made about the link between hope and activism is that participants from both sides explicitly mentioned that attending binational activities heightens their expectations for peace. Seeing people from the other side come together to share concerns and express solidarity increases the belief that peace is possible. Through these activities, friendships form across borders and religions. So, in a way, binational activism provides proof of the feasibility of peace, a small-scale version of peace between nations and societies. Perhaps those who do not attend these activities cannot fully grasp these insights. In this regard, activists have unique experiences concerning the feasibility of peace: they have seen it with their own eyes. Experiencing these "peace realities" offers a glimpse of what peace can look like, a reality that non-activists cannot see.

This point relates to the interruptions we noticed in the presumed reciprocal relations between activism and hope. Our interviewees testified that their activism leads to increased hope (i.e., greater expectations that peace can come about) but that these expectations do not necessarily translate into increased motivation to act for peace. Most of the participants in our study dismissed the idea that expectations for peace propelled their activism. Why does the interruption occur?

As elaborated throughout the book, hope is often regarded as essential to well-being, something all humans need (e.g., Frankl, 1946; Fromm, 1968, Hume, 1740/2003; Tillich, 1965). Activists are no exception. The idea that hope is a basic human need was articulated by the activists themselves. Linking the dots, we can cautiously infer that peace activists working in the discouraging conditions of intergroup conflict and animosity are active, at least in part, to nurture and replenish their hopes, especially when these hopes start dwindling. In other words, hope becomes the end, not only the means of activism. Amjed's remark (16) that discouragement motivates him to act for peace makes sense when we adopt this interpretation. He, and possibly other peacebuilders, are looking forward to the activities to refill their hopes, which are constantly depleted by the destruction and suffering caused by the violent conflict. Understanding hope as an end, not only as a means, is found in philosophical (e.g., Mill, 1875) and psychological approaches to hope (e.g., Breznitz, 1986). Exploring the testimonies of peace activists suggests a similar phenomenon. In sum, peacebuilding activities are directed to some degree to replenish activists' hope, not only to achieve peace.

Before we continue, we wish to address the question of power relations within these organizations. Groups locked in an asymmetrical dispute have very different experiences of conflict and war (Maoz & McCauley, 2008; Rumelili & Çelik, 2017). The disparities are evident in the conflict in Israel-Palestine, where power differences are stark. Israelis live in relatively benign conditions of conflict, while Palestinians' experiences as an occupied (in the West Bank) or besieged (in the Gaza Strip) society are incredibly harsh. It is thus not surprising that studies conducted on both groups found substantial gaps in Palestinians' and Israelis' attitudes (Maoz, 2000; Zeitzoff, 2018) and hope for peace (Leshem & Halperin, 2020a). Though not a part of our initial hypotheses, we looked for disparities between Palestinian and Israeli accounts of hope.

To our surprise, we could not detect stark differences in how Israelis and Palestinians talked about their hopes and activism. Of course, the examples given by activists from each nationality were very different, and their unique experiences of conflict were manifested in the interviews. Nonetheless, it was hard to identify substantial disparities between Israeli and Palestinian activists regarding the issues of hope, activism, and their connection. We propose two explanations for the similarity in the accounts of Israeli and Palestinian interviewees. The first is that the binational setting of the four NGOs provides, as much as possible, an equal standing for Israeli and Palestinian members. The binational NGOs offer a rare opportunity to experience and exercise partnerships that may minimize the differences between the activists, at least when they talk about peacebuilding and hope. Second, it seems that Israeli activists are well aware of power asymmetry and acknowledge that most of the responsibility for reaching peace lies in the hand of Israel. Therefore, the hopes of both Israeli and Palestinian activists were mostly, though not exclusively, directed at the need to change the attitudes and behaviors of Israeli society toward Palestinians and their struggle for self-determination. This consensus attenuated potential differences and possibly contributed to the resemblance between Israeli and Palestinian interviews.

This chapter contributes to the study of hope, conflicts, and their relationships in several ways. First, we reveal that peace activists are driven more by their uncompromising desires for peace rather than their expectations that peace is possible. Chapter 7 also suggests that the wish dimension is often the more robust predictor of the two when it comes to peace-promoting outcomes. Therefore, our qualitative analysis supports the quantitative models that explore the predictive power of each dimension. Another contribution is our proposition that, in some situations, political activism is directed at replenishing hope rather than achieving political change.

Some methodological advancements accompany these theoretical contributions. First, the study complements quantitative investigations of hope in conflicts by providing broad contexts and examining how members of two rival societies understand and interpret hope. Furthermore, the current study pioneers the qualitative research of hope based on the bidimensional model. The bidimensional approach proved fruitful in quantitative studies on hope (e.g., Leshem & Halperin, 2021). This chapter provides initial evidence for the utility of the bidimensional model in qualitative research. First and foremost, we believe that, without distinguishing between the two dimensions, it was impossible to decipher activists' use of the word "hope" as sometimes they referred to "hope" to express their desires for peace and sometimes to express their expectations that peace will materialize. Adopting the binational model makes interpretation straightforward. The fact that the two dimensions of hope were readily detected in the corpus hints at the applicability of the bidimensional model in qualitative research. This and more. The qualitative use of the bidimensional model helped to pinpoint which dimension drives peace activists working during conflicts and political turmoil and which is less influential in their work.

Several limitations of the study should be mentioned. First, our decision to focus only on binational NGOs might confine our ability to generalize our findings to other types of peace organizations. It could certainly be the case that members of NGOs working within each community have different interpretations and experiences of hope. The small number of female Palestinian interviewees is also a limitation. We aim to enlarge the sample of activists and investigate activists from uni-national NGOs in the future. Finally, it will be instructive to conduct similar studies among activists in other conflict zones. The qualitative strand of the global Hope Map Project might be a promising avenue by which to pursue a qualitative approach to understanding hope amidst conflict.

Conclusion

What are the broader implications of this study? The first is that the role of hope in political processes should not be taken for granted or at face value. Demonstrated here at the grassroots level, it seems that, as a driver of political change, hope has less to do with believing that change is possible and more to do with believing that change is essential. This insight resonates with Vaclav Haval's comments that hope has nothing to do with prognosis (Havel, 1990). Havel, who spent years in jail for opposing the Czechoslovakian government, talked about hope as a driver of his choice to dissent, not hope as an expectation but as a wish.

The second is that activism is often aimed at maintaining hope for political transformation, not only achieving the transformation. Stated differently, *hope* for change (in this case, peace) becomes the target of the behavior, not the change itself. Readers may look critically at the finding. Peacebuilding activities are supposed to bring about peace, not only the hope for it. However, we think it is safe to say that peacebuilders, like the twenty interviewed in this study, genuinely seek political change in the form of a just and sustainable peace. Their commitment and enthusiasm are admirable and serve as an inspiration. Analyzing the interviews, it is clear that their need to replenish hope does not come at the expense of seeking real change. Working on the frontline of peacebuilding is demanding. Even peacebuilders can use a little hope.

References

Averill, J., & Sundararajan, L. (2005). Hope as a rhetoric: Cultural narratives of wishing and coping. In J. Eliott (Ed.), *Interdisciplinary Perspectives on Hope* (pp. 133–166). Nova Science.

Braun, V., & Clarke, V. (2006). Using thematic analysis in psychology. *Qualitative Research in Psychology*, *3*(2), 77–101. https://doi.org/10.1191/1478088706qp063oa

Breznitz, S. (1986). The effect of hope on coping with stress. In M. H. Appley (Ed.), *Dynamics of Stress* (pp. 295–306). Plenum.

Burger, M. J., Hendriks, M., Pleeging, E., & van Ours, J. C. (2020). The joy of lottery play: Evidence from a field experiment. *Experimental Economics*, *23*(4), 1235–1256. https://doi.org/10.1007/s10683-020-09649-9

Cohen-Chen, S., & Van Zomeren, M. (2018). Yes we can? Group efficacy beliefs predict collective action, but only when hope is high. *Journal of Experimental Social Psychology*, *77*, 50–59. https://doi.org/10.1016/j.jesp.2018.03.016

Courville, S., & Piper, N. (2004). Harnessing hope through NGO activism. *Annals of the American Academy of Political and Social Science*, *592*(1), 39–61. https://doi.org/10.1177/0002716203261940

Dowty, A. (2006). Despair is not enough: Violence, attitudinal change, and "ripeness" in the Israeli-Palestinian conflict. *Cooperation and Conflict*, *41*(1), 5–29. https://doi.org/10.1177/0010836706060930

Frankl, V. (1946). *Man's search for meaning*. Rider.

Fromm, E. (1968). *The revolution of hope*. Harper & Row.

Furlong, C., & Vignoles, V. L. (2021). Social identification in collective climate activism: Predicting participation in the environmental movement, extinction rebellion. *Identity*, *21*(1), 20–35. https://doi.org/10.1080/15283488.2020.1856664

Greenaway, K. H., Cichocka, A., van Veelen, R., Likki, T., & Branscombe, N. R. (2016). Feeling hopeful inspires support for social change. *Political Psychology*, *37*(1), 89–107. https://doi.org/10.1111/pops.12225

Havel, V. (1990). *Disturbing the peace: A conversation with Karel Hvížďala*. Vintage.

Hermann, T. (2009). *The Israeli peace movement: A shattered dream*. Cambridge University Press.

Hume, D. (1740/2003). *A treatise of human nature*. Courier.

Jamal, A., & Lavie, N. (2021). Hope and creative work in conflict zones: Theoretical insights from Israel. *Sociology*, 00380385211056218. https://doi.org/10.1177/00380385211056218

Kant, I. (1781). *Immanuel Kant's critique of pure reason*. McMillian.

Lazarus, R. (1999). Hope: An emotion and a vital coping resource against despair. *Social Research*, *66*(2), 653–678. https://www.jstor.org/stable/40971343

Leshem, O. A. (2017). What you wish for is not what you expect: Measuring hope for peace during intractable conflicts. *International Journal of Intercultural Relations*, *60*, 60–66. https://doi.org/10.1016/j.ijintrel.2017.06.005

Leshem, O. A. (2019). The pivotal role of the enemy in inducing hope for peace. *Political Studies*, *67*(3), 693–711. https://doi.org/10.1177/0032321718797920

Leshem, O. A., & Halperin, E. (2020a). Hope during conflict. In S. C. van den Heuvel (Ed.), *Historical and multidisciplinary perspectives on hope* (pp. 179–196). Springer International. https://doi.org/10.1007/978-3-030-46489-9_10

Leshem, O. A., & Halperin, E. (2020b). Hoping for peace during protracted conflict: Citizens' hope is based on inaccurate appraisals of their adversary's hope for peace. *Journal of Conflict Resolution*, *64*(7–8), 1390–1417. https://doi.org/10.1177/0022002719896406

Leshem, O. A., & Halperin, E. (2021). Threatened by the worst but hoping for the best: Unraveling the relationship between threat, hope, and public opinion during conflict. *Political Behavior*, *20*. https://doi.org/10.1007/s11109-021-09729-3

Litt, M. D., Tennen, H., Affleck, G., & Klock, S. (1992). Coping and cognitive factors in adaptation to in vitro fertilization failure. *Journal of Behavioral Medicine, 15*, 171–187.

Maoz, I. (2000). An experiment in peace: Reconciliation-aimed workshops of Jewish-Israeli and Palestinian youth. *Journal of Peace Research, 37*(6), 721–736. https://doi.org/10.1177/0022343300037006004

Maoz, I., & McCauley, C. (2008). Threat, dehumanization, and support for retaliatory aggressive policies in asymmetric conflict. *Journal of Conflict Resolution, 52*(1), 93–116. https://doi.org/10.1177/0022002707308597

Mill, J. S. (1875). *Theism* (R. Taylor, Ed.).

Mills, J., & Birks, M. (2014). *Qualitative methodology: A practical guide*. SAGE.

Nardini, G., Rank-Christman, T., Bublitz, M. G., Cross, S. N. N., & Peracchio, L. A. (2021). Together we rise: How social movements succeed. *Journal of Consumer Psychology, 31*(1), 112–145. https://doi.org/10.1002/jcpy.1201

Perry, D. (2011). *The Israeli-Palestinian peace movement: Combatants for peace*. Springer.

Rumelili, B., & Çelik, A. B. (2017). Ontological insecurity in asymmetric conflicts: Reflections on agonistic peace in Turkey's Kurdish issue. *Security Dialogue, 48*(4), 279–296. https://doi.org/10.1177/0967010617695715

Rynhold, J., & Waxman, D. (2008). Ideological change and Israel's disengagement from Gaza. *Political Science Quarterly, 123*(1), 11–37.

Schwebel, M. (2008). Peace activists: Maintaining morale. *Peace and Conflict: Journal of Peace Psychology, 14*(2), 215–224. https://doi.org/10.1080/10781910802017370

Shulman, D. (2007). *Dark hope: Working for peace in Israel and Palestine*. University of Chicago Press.

Small, M. (1988). *Johnson, Nixon, and the doves*. Rutgers University Press.

Thomas, E. F., Mavor, K. I., & McGarty, C. (2012). Social identities facilitate and encapsulate action-relevant constructs: A test of the social identity model of collective action. *Group Processes & Intergroup Relations, 15*(1), 75–88. https://doi.org/10.1177/1368430211413619

Tillich, P. (1965). *The right to hope*. Harvard Divinity School.

van Zomeren, M. (2013). Four core social-psychological motivations to undertake collective action. *Social and Personality Psychology Compass, 7*(6), 378–388. https://doi.org/10.1111/spc3.12031

van Zomeren, M., Pauls, I. L., & Cohen-Chen, S. (2019). Is hope good for motivating collective action in the context of climate change? Differentiating hope's emotion- and problem-focused coping functions. *Global Environmental Change, 58*, 101915. https://doi.org/10.1016/j.gloenvcha.2019.04.003

Wlodarczyk, A., Basabe, N., Páez, D., & Zumeta, L. (2017). Hope and anger as mediators between collective action frames and participation in collective mobilization: The case of 15-M. *Journal of Social and Political Psychology, 5*(1), 200–223. https://doi.org/10.5964/jspp.v5i1.471

Zeitzoff, T. (2018). Anger, legacies of violence, and group conflict: An experiment in post-riot Acre, Israel. *Conflict Management & Peace Science, 35*(4), 402–423. https://doi.org/10.1177/0738894216647901

6
The Politics of Hope and the Politics of Skepticism

Oded Adomi Leshem, Ilana Ushomirsky, Emma Paul, and Eran Halperin

> A leader is a dealer in hope.
>
> **—Napoleon Bonaparte**

On a freezing morning on February 24, 2022, Russian troops invaded Ukraine. A month after the invasion, the death toll exceeded 10,000.[1] The bravery of the Ukrainian people facing a military force much stronger than theirs has captured the hearts and minds of many around the world. But perhaps the most captivating figure that emerged in the first days of the fighting is that of Ukraine President Volodymyr Zelenskyy. Zelenskyy's simple and charismatic nature and bold steadfastness in the face of Russian aggression gained him immediate international respect. Addressing the British House of Commons two weeks into the invasion, Zelenskyy said,

> We will not give up, and we will not lose. We will fight until the end, at sea, in the air. We will continue fighting for our land, whatever the cost. We will fight in the forests, in the fields, on the shores, in the streets.

Well known to his audience, Zelenskyy's words resonated with Churchill's famous speech given in the same venue in 1940 after the fall of Dunkirk.

> I have, myself, full confidence that if all do their duty, if nothing is neglected, and if the best arrangements are made, as they are being made, we shall prove ourselves

[1] https://www.theguardian.com/world/live/2022/mar/24/russia-ukraine-war-latest-news-zelenskiy-expects-meaningful-steps-at-nato-eu-and-g7-summits-live?page=with:block-623be1038f08118734a71322

Hope Amidst Conflict. Oded Adomi Leshem, Ilana Ushomirsky, Emma Paul, and Eran Halperin Oxford University Press.
 DOI: 10.1093/oso/9780197685303.003.0006

> once again able to defend our Island home, to ride out the storm of war, and to outlive the menace of tyranny. . . . We shall fight on the beaches, we shall fight on the landing grounds, we shall fight in the fields and in the streets.

The deliberate similarity of words was aimed to boost the audience's empathy toward the Ukraine catastrophe. Most relevant to this book's exploration, both speeches reflect a political approach termed the "politics of hope" (Botwinick, 2010; Rorty, 1999). The politics of hope is the politics of will, vision, efficacy, and promise. The content of these future-oriented constructs may greatly vary from one context to another, but the politics of hope, like hope itself, is about the desires and the possibilities of the future.

The desires to attain a political goal and the belief in the possibility of attaining that goal seem to be innate to politics and leadership. However, and perhaps counterintuitively, skepticism can also be an effective political approach. Politics of skepticism keeps dreams to a minimum and instead focuses on maintaining the present and preventing future losses. Instead of promises for the future, the politics of skepticism is about the here and now. A good example of the politics of skepticism is US Congresswoman M. T. Greene's reaction to Zelenskyy's request for American aid. "We cannot fund more of it by sending money and weaponry to Ukraine to fight a war they cannot possibly win. . . . It's not our responsibility to give President Zelenskyy and the Ukrainian people false hope about a war they cannot win."[2]

Greene's skeptical words stand in contrast to Zelenskyy's (and Churchill's) rhetoric of hope but are nevertheless persuasive. Surely, as of April 2022 (the time of writing this chapter), it was impossible to know how the war in Ukraine would unfold, and "Prophecy," as the Talmud says, "was given to the fools and infants." It could be the case that Ukrainians' hope to drive out the Russian forces would be completely fulfilled. Alternatively, these hopes may be dashed in never-ending Russian control of Ukrainian land and politics. Who is to know?

When leaders employ politics of hope, they prophesize and promise a better future in the image of victory or prosperity. The politics of skepticism makes no commitment (Botwinick, 2010). When leaders practice the politics of skepticism, they downplay dreams and promises in the name of prudence and reason. Statespersons claiming to speak in the name of reason may be honest or manipulative. What is essential is that anticipations for a better future are kept at a minimum, leaving more attention to immediate governing

[2] Marjorie Taylor Greene for Congress (March 16, 2022). The war in Ukraine. Retrieved from https://www.youtube.com/watch?v=SEKwWgbva3U

needs. Indeed, the politics of skepticism lacks the attractive aura of hope. It is a grayer type of politics, one without glory or allure. Yet, when used effectively, the politics of skepticism can be an extremely efficient mobilizing strategy (Navot et al., 2017).

Is there a connection between conflict dynamics and the politics of hope and skepticism? Up until this point, the book focused on hope among the masses. It explored the antecedents of hope in large-scale samples and examined hope among grassroots activists. However, leaders and statespersons might have the most influence on the dynamics and trajectories of conflict. Understanding how elites' hope (or the lack of hope) for peace play out in the context of conflict is likely to be instrumental in studying the connections between conflict and hope.

Written with Ilana Ushomirsky, Emma Paul, and Eran Halperin, this chapter presents a theoretical and empirical link between the two political approaches and the dynamics of intractable asymmetrical conflicts. More specifically, we examine how hope and skepticism about peace are expressed by Israeli and Palestinian leadership and tie these expressions to the power structure of the conflict.[3] We begin by laying out the theoretical background of the politics of hope and the politics of skepticism while providing historical examples of the two approaches. We then describe the two political approaches in Israeli and Palestinian leadership, which culminates in a provocative hypothesis: that Israeli leaders have adopted the politics of skepticism when it comes to conflict and peace, whereas Palestinian leadership has adopted the politics of hope. We then test this claim by analyzing a corpus of forty-six speeches made by Israeli and Palestinian representatives in the United Nations General Assembly. We offer several explanations of our findings at the end of the chapter, as well as suggestions for further explorations of the politics of hope and skepticism.

The Politics of Hope

The politics of hope builds on the premise that social transformation is propelled by people's hopes (Bloch, 1959, 1970; Capps, 1968; Fromm, 1968; Rorty, 2002; Tillich, 1965). Whether the transformation led societies to establish liberal democracies or communist republics, humans' hopes, it is believed, made them transpire (Rorty, 1999). Politics of hope presupposes that political

[3] Consistent with the bidimensional conceptualization, I use the word "hope" to refer to the two dimensions at the same time. However, when referring to only one dimension I do not use "hope" but the specific dimension: namely, "wishes" or "expectations."

change, from the emancipation of nations to the abolishment of slavery, was made possible because the masses and their leaders willed them and believed these changes can materialize.

A fundamental assumption underlining the politics of hope is that the future of humankind can, and therefore should be molded by plans and actions in the present. Jacques Derrida, for example, saw humanity as an ideal not yet achieved (Russell, 2001). He further claimed that humans must work in the present to fulfill this ideal in the future. Despite pressures from the state and other institutions that strive to limit the aspirations and faith of the people, man can and thus must seek the ideal (Russell, 2001). The necessity of being an idealist and working to live up to this ideal was also advocated by Victor Frankl, who criticized the realist approach that seeks to understand humans as they "really" are. Building on Goethe, Frankl believed humans should be understood as they "should be" in order to make them "what they can be."

Victor Frankl, Eric Fromm, Ernest Bloch, Richard Rorty, and others are often considered the leading voices of the "philosophy of hope movement" that gained momentum in the second half of the twentieth century (Capps, 1968). For these "hope philosophers," politics is an act of hope. "Political deliberation presupposes hope," writes Rorty, "it presupposes that things can be changed for the better" (Rorty, 2002, p. 152). Bloch considered hope to be a political endeavor by its very nature and urged humankind to "plunge beyond the horizon in that very difficult sphere of reality . . . not the reality which is being present, not the reality in process, but the reality of the not yet" (Bloch, 1970, p. 63).

Two points are crucial for understanding the politics of hope. The first is that political hope is linked to human suffering (Eckardt, 1982; Fromm, 1966). Fromm noted that hoping for political change can only stem from hardship, not comfort. When people are content with their lives, when the status quo produces profits, hope for political transformation will be absent. One hopes for freedom, justice, or peace only when one suffers from repression, injustice, or war. Fromm continued his argument by saying that "the beginning of liberation lies in man's capacity to suffer. . . . The suffering moves him to act against the oppressors, to seek the end of the oppression. . . . If man has lost the capacity to suffer, he also has lost the capacity for change" (Fromm, 1966, p. 87).

The second point is that the politics of hope is directly connected to rhetoric and discourse. Hope is an appealing and effective tool of persuasion because it assumes human agency and ability to predict and shape the future. As a rhetorical tool, expressions of hope convey courage, strength, and resolve (recall Zelenskyy's and Churchill's speeches). Hope is thus attractive to audiences that urgently need a positive outlook to cling to. Politicians have long understood that hope captivates and mobilizes constituents and therefore "talked" hope to the masses. Jessie Jackson's "Keep hope alive," MLK's "I have a dream," and Barak Obama's "Yes we can" are clear examples of the rhetoric of hope.

The politics and rhetoric of hope can be found across the political spectrum and in various countries. For instance, political parties use the appealing idea of hope by including it in their names. The right-wing Israeli party "New Hope," the center-left Guatemalan party "National Union for Hope," the liberal party "Hope" from Slovakia, and parties in countries like Japan, Nigeria, and Lebanon, have drawn on the positive and motivational connotations of hope when forming their party's name and agenda. Furthermore, some highly successful political leaders have utilized hope throughout their campaigns and careers; Barack Obama is one of the best known. In addition to his famous "Yes we can" slogan, Obama's speeches and writing were infused with the rhetoric of hope. For example, drawing on American history and stories of shared suffering, he reminded Americans of the uniting force of hope in his 2004 address at the Democratic National Convention.

> In the end, that's what this election is about. Do we participate in a politics of cynicism, or do we participate in a politics of hope? . . . I'm not talking about blind optimism here. . . . I'm talking about something more substantial. It's the hope of slaves sitting around a fire singing freedom songs; the hope of immigrants setting out for distant shores. . . . Hope—Hope in the face of difficulty. Hope in the face of uncertainty. The audacity of hope! In the end, that is God's greatest gift to us, the bedrock of this nation.

Two years later, setting the tone for his upcoming candidacy, Obama released his second book, *The Audacity of Hope*. Expanding on his core values and visions for US politics, he reiterated his belief in the need for a politics of hope, reminding his readers of what they share: "Common hopes, common dreams, a bond that will not break." Obama continued to draw on the politics of hope throughout his campaign for the 2008 presidency. In his inaugural speech as President of the United States, amid the financial crisis of 2008–2009, Obama's rhetoric of hope remained unshaken as he declared: "On this day, we gather because we have chosen hope over fear, unity of purpose over conflict and discord."

Nelson Mandela's speeches provide another example of the rhetoric of hope. In his orations, Mandela stressed the power of hope in the face of adversity and as a motivating force. His inauguration speech as State President in 1994 displayed his fundamentally optimistic vision for the new South African democratic state. "We have triumphed in the effort to implant hope in the breasts of the millions of our people. We enter into a covenant that we shall build the society in which all South Africans, both black and white, will be able to walk tall, without any fear in their hearts, assured of their inalienable right to human dignity—a rainbow nation at peace with itself and the world." Similarly, former President of Liberia Ellen Johnson-Sirleaf described the significance of hope for herself personally and for the success of Liberia's democracy. "In my journey, I have come to value hope and resilience. As an actor in Liberia's history, as it has unfolded over the last forty years, I have seen these characteristics come full circle."

The notions that hope is a catalyst of political change and a powerful mobilization tool might be regarded as simple common sense, perhaps too obvious to mention. However, the politics of hope is contested by the less known and perhaps less intuitive idea of the politics of skepticism.

The Politics of Skepticism

Political skepticism can be traced back to skeptical philosophical movements in ancient Greece and Rome and the writings of Montaigne, Hume, Kant, and Nietzsche (Laursen, 1992; Shaw, 2007). Philosophical skepticism questions the possibility of knowledge because all knowledge is subjectively conceptualized, categorized, and processed in the human mind (Botwinick, 2010; Laursen, 1992). In its purest form, skepticism argues that knowledge is impossible, ultimately freeing humans from holding views, making judgments, and executing choices. A less radical approach to skepticism encourages doubt toward the world of knowledge and suspicion toward certainty and definiteness.

Political skepticism is built on skeptical philosophy. It shifts the attention from pursuing political visions of the unknown future to the practical maintenance of the more certain political present. Contemporary writings on political skepticism are perhaps associated most with Michael Oakeshott, the twentieth-century English thinker who wrote extensively on why skepticism is the preferred system of politics (Botwinick, 2010; Gamble, 2021b; Oakeshott, 1996; Tseng, 2013). According to Oakeshott, the government should be understood by citizens and leaders alike as the apparatus needed to maintain immediate order (Navot et al., 2017). Governments, claimed Oakeshott, should not be a tool for advancing future well-being because any plan for the future is deceiving and thus futile.

The term "politics of skepticism" also includes a skeptical stance toward constituents' aspirations for the future (Gamble, 2021b). Citizens have desires and passions. This is natural. Yet, to the extent that it seeks to function properly, a good government must remain skeptical of these aspirations, especially those directed at political change. Oakeshott claims that a significant advantage of political skepticism is that it enables efficient governing by encouraging citizens to lower their demands for change and reform. When demands for reform are kept at a minimum, leaders and civil servants have the needed leeway to "do their job" by providing quality administration and pragmatic governance.

Pragmatic governance is one of the cornerstones of political skepticism (Canovan, 1999). However, apart from streamlining the practicalities of governance, political skepticism has another advantage: it counterbalances peoples' delusions of unattainable political goals (Gamble, 2021b). By dampening passions for unattainable goals, citizens are freed to pursue their desires in other aspects of their lives, be they vocational, economic, or spiritual. Overall, political skepticism is intended to improve the function of the state and the lives of its citizens by muting peoples' futile strivings to steer society toward a particular political vision.

At this stage, two distinctions should be made between political skepticism and other related concepts. First, political skepticism does not necessarily overlap with political conservatism (Gamble, 2021b). Pure political skeptics posit that deliberation and planning should be kept to a minimum lest they come at the expense of immediate governmental duties (Botwinick, 2010). Conservatives, like liberals, seek to actively shape the future. By reinstating traditional ideas and deliberating how these ideas could be endorsed by generations to come, conservatives are certainly not passive toward the future. Second, political skepticism should not be conflated with political ideology (Botwinick, 2010). The fact that political skeptics discourage public aspirations for a better future does not mean that they believe leadership should fill that space. Both totalitarian and democratic ideologies offer a vision for the future (Rorty, 1999). For instance, many totalitarian regimes, be they communist or fascist, manufacture a vivid picture of an ideal future and then actively advocate, sometimes coerce, the populace to endorse it. Liberal democrats also actively fuel citizens' passion with a particular image of the future. Political skepticism, which seeks to detach politics from the future, is thus at odds with both totalitarian and liberal democratic thought.

Turning to compare political skepticism with hope, or more correctly, lack thereof, it seems clear that political skepticism negates both dimensions of hope. First, political skepticism aims to reduce the aspirations and desires for political transformation. In this sense, political skepticism is antagonistic to the wish dimension of hope. At the same time, political skepticism proposes that governments should not promise to deliver future change (Gamble, 2021a). The expectations dimension of hope is therefore also antithetical to the politics of skeptics. Both the wish to attain political goals and the expectations of attaining them are dismissed from skeptics' point of view. Overall, as an approach that minimizes deliberation of the future, political skepticism is the opposite of political hope.

At first glance, it might appear that political skepticism is destined for eternal unpopularity. What politician would invest in a political approach that does not promise a better future? Who would follow a leader who does not offer a vision and a pledge to fulfill it? The idea of political skepticism seems to contradict what many believe is a simple truism: that hope for social advancement is the motor of all politics. However, a more thorough examination reveals skepticism's attractive features. Skeptical leaders offer a seemingly realistic and unbiased account of life. They say things "as they are" without sugar-coating or making unreliable promises. As such, skeptical politicians may appear less naïve, more mature, and tough. They urge their followers to divert their energies to the here and now, instead of focusing on delusional passions for irrelevant futures. Strong politicians should not be concerned with dreams, argues Oakeshott, but with the ability to accept that reality may be distasteful, disagreeable, or morally obnoxious (Gamble, 2021b).

Examples of politicians practicing political skepticism are less prevalent, perhaps because optimism and hope seem much more appealing. Nevertheless, some interesting examples of the politics of skepticism might be found in US politics. For example, President George H. W. Bush's leadership style has been described as "Oakeshottian" due to Bush's emphasis on managing the present rather than planning for the future (Hammer, 1995). The skeptical stance of President Bush is perhaps best manifested in his reluctance to articulate a coherent and clear vision for his political party and the United States more generally (Mullins & Wildavsky, 1991). In line with the concept of political skepticism, Rose (1991) has argued that Bush was a "guardian president" as he sought to safeguard what already existed rather than change the direction of government. Moreover, Bush's preference to solve problems as they came instead of anticipating their emergence provides more evidence for his tendency for political skepticism. As disclosed by one of his staff members: "one of my biggest disappointments was that nobody [thought] beyond next Wednesday" (Hammer, 1995, p. 306).

A more extreme version of political skepticism can be found in the public statements of Andrew Yang, one of the candidates for the Democratic Party's Presidential Primaries of 2020. Characteristic of the politics of skepticism, Yang stressed, "We have to start dealing with the world as it is." The statement relates to the dangers of hope as a path that can lead to delusive dreams and inevitable disappointments. Yang was consistent in his politics of skepticism when commenting on concrete policies. When asked about climate change during a CNN debate, he delivered a gloomy message: "This is going to be a tough truth, but we are too late . . . we are ten years too late." While this claim was criticized for being fatalistic, it could be argued that Yang was simply being honest. Yang continued his skeptical approach regarding the COVID-19 pandemic, saying that he was worried about optimistic politicians being unaware of the gravity of the crisis. Yang's skeptical approach was often labeled as "bad politics."[4] Still, one must also ask whether this negative label stems from humans' preference for optimistic dreams over unpleasant truths.

The Politics of Hope and the Politics of Skepticism in Palestine and Israel

We now return to the context of conflict and offer a brief historical account concerning the politics of hope and the politics of skepticism among Palestinian and Israeli leaders. This account serves as an introduction to the following subsection, in which we systematically compare the two approaches in speeches of Palestinian and Israeli leaders. Our main argument in the introduction is that, since the final years of the 1990s, the Israeli leadership has been applying the politics of skepticism, whereas the politics of hope has been more commonly and consistently practiced among Palestinian leaders. At this point, we note that existing literature about this topic is scant. Apart from Navot et al. (2017), we know of no study that examined hope and skepticism in Israeli or Palestinian leadership. We start by describing the two political approaches in Israel.

The Politics of Hope and the Politics of Skepticism in Israel

By and large, the Jewish Israeli society can be characterized as a hopeful one with a strong sense of agency. After all, building a viable and prosperous state from the ashes of the Holocaust requires high aspirations and a firm belief in attaining these aspirations (Parsons & Schneider, 1974). However, as

[4] https://grist.org/article/andrew-yang-fearmonger-or-climate-realist/;%20https:/www.theatlantic.com/science/archive/2019/08/andrew-yangs-horrific-debate-answer-climate-change/595267/

we outline in the following paragraphs, Israelis' hopefulness is much more equivocal and unstable regarding peace. We further suggest that there has been a shift in Israelis' hope for peace around the turn of the century and that this shift can be detected by utilizing the bidimensional approach to hope. Therefore, in our brief historical introduction to Israeli leaders' expressions of hope, we will break hope into two dimensions: the wish for peace and the expectations that peace will someday be achieved.

Until the end of the past millennia, the expectations for peace among the Israeli leadership fluctuated depending on the circumstances. For example, during tense periods, Israeli leadership adopted a skeptical stance about the possibility of peace. Skepticism was expressed, for instance, by Abba Eban, Israel's Minister for Foreign Affairs during the late 1960s and early 1970s, in his response to the declarations made by eight Arab countries in September 1967. Responding to the Arab's "Three No's" ("No recognition of Israel, No negotiation with Israel, No peace with Israel"), Eban said that the declaration proves how far we are from peace. "[The Arab's] declaration shows stagnation in the thinking of the Arabs and a stagnation in the new situation in the Middle East."[5] Four years later, Menahem Begin expressed similar skepticism about the chances of future peace in his response to an UN-led attempt to bring Egypt and Israel to the negotiation table. In an op-ed published in a popular Israeli newspaper, Begin wrote: "The treaty between the Arabs and Israel is not getting nearer, not advancing, not even in a micro-millimeter."[6]

However, when the prospects of peace seemed to be nearing, such as when Egypt was signaling its desire to negotiate, the leadership in Israel adopted a more optimistic stance. For example, six years after his skeptical op-ed, Menahem Begin, now Israel's Prime Minister, conveyed his high expectations for peace when speaking to the Israeli Parliament during Sadat's visit to Jerusalem. Turning to Sadat, Begin said, "Let us negotiate a peace accord, Mister President, as free people. And God willing, and this is our honest belief, there will come a day when we sign it. . . . Then we will know that war is over. We extended our hand for peace, we shook hands as comrades, and there will be a prosperous future for all nations of the region."[7]

The overall picture is that, quite naturally, levels of expectations for peace change in response to unfolding events in the region. When Arab leaders signaled their aspirations for peace, the expectation dimension of hope increased among the Israeli leadership. When Arab leaders signaled antagonism or

5 https://www.nli.org.il/he/newspapers/mar/1967/09/10/01
6 https://db.begincenter.org.il
7 https://www.knesset.gov.il/process/docs/beginspeech.htm

when tensions and hostilities were on the rise (as is the case most of the time), Israeli leaders expressed low expectations for peace. Relying on cues conveyed by the rival party to assess the likelihood of peace is a reasonably valid process. However, it also shows a lack of responsibility and agency because it assumes that the chances of achieving peace lie chiefly in the hands of the adversary (Leshem & Halperin, 2020b).

In contrast to the oscillating expressions of expectations for peace, Israeli leaders' expressions of the *desire* for peace (hope's wish dimension) were less equivocal and more stable until the turn of the millennium. Israel's 1948 declaration of independence contains the first expression of the state's formal endorsement of the wish for peace. Since then, the desire for peace with the Arab nations surrounding Israel has been professed in Israeli leaders' statements throughout the first decades of Israel's independence (Oren, 2019). One of the most emotional expressions of Israel's yearning for peace was delivered by Menachem Begin, the Israeli Prime Minister who ultimately negotiated the peace agreement with Egypt. "We are fighting today for peace, and it is bliss that we are in this moment," Begin said to Israeli Parliament Members who opposed the peace accord, "Yes, peace has hardships, peace has pain, peace has casualties; but they are all belittled by the casualties of war." When accepting the Nobel Prize for Peace, he added: "Peace is the beauty of life, the rising of the sun, the smile of a child . . . the progress of man, the victory of truth. Peace is all this, and more, and more, and more."

Unsurprisingly, formal declarations of Israel's desires for peace peaked in the first half of the 1990s, when the peace agreement with Jordan was signed. Similar declarations were made when Israel's attention was moving from the neighboring Arab states to the option of peace with the Palestinians. "Let us all turn bullets to ballots, guns to shovels," said Israel's Foreign Minister Shimon Peres at the signing of the declaration of principles with the Palestinians. "As we have promised, we shall negotiate with you a permanent settlement and with all our neighbors a comprehensive peace. Peace for all," Peres declared. Perhaps the most iconic moment of the politics of hope was when Prime Minister Itzhak Rabin joined the chants of tens of thousands gathered in Tel-Aviv in 1995 to support the Oslo Peace Process. The demonstrators sang an old Israeli tune called "A Song for Peace." "You better sing the song for peace with all your might," sang Rabin along with the crowd. Several minutes later, he was shot.

Until Rabin's murder, the expectations for peace depended on signals from the Arabs while the wishes for peace were more independent, stable, and unquestionable, part of the ethos of Jewish Israeli society (Oren, 2019). The assassination of Rabin, perhaps the chief bearer of hope at the time, may have been the trigger for Israel's shift from the politics of hope to the politics of skepticism. Other tragic events occurring in the second half of the 1990s have helped the transition to the politics of skepticism. For example, a few months after Rabin's assassination, sixty Israeli civilians were killed in four consecutive attacks executed by Palestinian suicide bombers. Several months later, Netanyahu was elected for his first tenure as Prime Minister on the promise that he alone would regain Israelis' sense of security. The downturn in hope for peace was accelerated with the collapse of the Camp David Peace Summit in July of 2000 and the eruption of the Second Intifada in September of the same year.

It is within this context that Ehud Barak, Prime Minister of Israel from 1999 to 2001, made his skeptical declaration that peace with the Palestinians is impossible. Several months after the failure of the Camp David Summit, Barak said, "the picture that is emerging, is that there is apparently no partner for peace. This truth is a painful one, but it is the truth, and we must confront it with open eyes and draw the necessary conclusions." The "no partner" paradigm quickly caught on and had a profound impact on Israelis' declining expectations for peace. The Israeli politics of skepticism had received its first iconic catchphrase: "There is no partner for peace."

Netanyahu's administration brought the politics of skepticism to higher levels. With Netanyahu, the "no partner" approach became a mantra.[8] Perhaps the best example of Netanyahu's politics of skepticism can be observed in his 2015 declaration that "Israelis will have to learn to live on their swords forever".[9] The declaration provides a pessimistic outlook on the possibility of peace. However, most importantly, it encourages the public to endorse it. At first glance, it might seem that the public would quickly reject messages like this. Why should people accept war as their permanent fate? Yet, as mentioned earlier, skepticism has its appeal as well. First, it addresses a basic human need for certainty and predictability (Fiske, 2010). "What has been will be again, and what has been done will be done again" reassures Koheleth in the book of Ecclesiastes (1:9). In the case of the Israeli-Palestinian conflict, predictability and certainty are gained when the all-too-familiar conflict is perceived to extend into the far future. Second, it frees people from the taxing, often frustrating desires for change and reduces the chances of disappointment. Last, a

[8] A noteworthy exception is Netanyahu's 2009 speech in Bar-Ilan University where he expressed his wishes for peace with the Palestinians.

[9] https://www.haaretz.com/israel-news/2015-10-26/ty-article/.premium/netanyahu-i-dont-want-a-binational-state-but-we-need-to-control-all-of-the-territory-for-the-foreseeable-future/0000017f-e67d-da9b-a1ff-ee7f93240000

skeptical stance portrays its conveyer as an honest "realist," a leader who says things "as they are" rather than selling cheap promises. Nothing will change, reassures Netanyahu. Better accept it than fight it.

A recent example of the widespread belief in the impossibility of peace is the success of a book by Micha Goodman, *Catch 67*. One of Goodman's key arguments is that resolving the Israeli-Palestinian conflict is impossible (Goodman, 2018). He thus encourages Israelis to mitigate the impact of the conflict rather than seek to resolve it. The Jewish Israeli public readily endorsed the book and its skeptical stance, which exonerates Israelis from striving for peace. The Hebrew version of the book became a major hit in Israel, leading the bestselling booklist for weeks. Goodman's skepticism about peace was welcomed not only among the Jewish Israeli public. Prime Minister Naftali Bennett and Minister of Justice Gideon Saar openly endorsed Goodman's skeptical approach.

However, the most alarming shift in the politics of hope in Israel is not the decline in the expectations for peace but in the wish for it. One of our arguments is that, once explicit and unequivocal, Israel's desires for peace have been fading. In addition to the shift in leaders' rhetoric, the decline in the desire for peace might be explained by the gradual disappearance of incentives for peace since the end of the Second Intifada. In terms of power relations between Israel and the Palestinians, Israel has had the upper hand since the end of the Second Intifada, and by a considerable margin. For example, since 2008, the average annual toll of Israeli casualties from conflict-related violence, civilian and military, has been approximately twenty-five.[10] When this number is compared to more than 350 annual Israeli casualties during the Second Intifada, the price might seem bearable on the national level.

In addition to the relatively low number of casualties (reflecting military superiority), Israel has also triumphed in the economic and diplomatic arena. Israel's economy has grown and become more resilient in the past two decades, attracting foreign investment in Israel's promising high-tech industry (Senor & Singer, 2011). On the diplomatic front, Israel has established an effective apparatus to prevent international pressure and curbed the sporadic attempts of any significant economic or cultural sanctions. Perhaps most remarkably, while Israel was tightening its hold on the West Bank, building Jewish Settlements, and scaling up violent measures against Palestinians in Gaza, it signed normalization treaties with Arab and Muslim countries such

[10] OCHA Data on Casualties (https://www.ochaopt.org/data/casualties).

as the United Arab Emirates, Bahrain, and Morocco. Overall, Israel has found a way to effectively manage the conflict (Zanany, 2018), consequently nullifying the need to solve it. Hope comes from suffering, argues Fromm (1966). Did the decrease in the extent that Israelis suffer from the conflict diminish their desires for peace?

Evidence from recently collected data sheds some light on this question. In 2017, Jewish Israelis' wish for peace (using the open-ended definition, see Chapter 4) was 5.05 (standard deviation [*SD*] = 1.28) on a scale from 1 to 6. In 2022, the score dropped to 4.52 (*SD* = 1.54). This drop is statistically significant ($t = 6.3, p < .001$). Looking deeper into the data, it seems that, in 2017, 50.5% reported that they wished for peace "very much." This percentage dropped to 35.8% in 2022. Of course, it is not that Israelis wish the conflict to continue. They simply lost their passion for peace.

Moreover, in a series of focus groups we conducted in Israel on the topic of conflict and peace, we noticed that peace was a nonissue for many participants. In this study, sixty-three Jewish Israelis from a wide range of political ideologies participated in a ninety-minute facilitated discussion on various issues related to the Israeli-Palestinian conflict. Before we revealed the session's topic, we first asked, "What are the important issues facing Israel at the moment?" Astonishingly, across the political spectrum, only one participant mentioned the conflict. Going back to elite politics, this indifference was perfectly reflected in the first UN address of the newly appointed Prime Minister Naftali Bennett on September 23, 2021. Speaking to the General Assembly, Bennett said, "Israelis don't wake up in the morning thinking about the conflict." It appears that Bennett's government fully adopted the politics of skepticism. Bennett and his foreign minister Yair Lapid reaffirmed the skeptical approach several months after the new government was sworn in. "We are not expecting any negotiations with the Palestinians," said Lapid, shutting the door for hope.

In sum, the politics of hope that characterized Israel until the mid-1990s was replaced by the politics of skepticism. Notably, the change is not only reflected in lower expectations for peace but also in lower wishes for peace. Several questions arise from this argument. The first is whether the politics of skepticism can be identified in Israeli leadership in a more systematic way rather than through the anecdotes presented thus far. The last part of this chapter tackles this question directly by quantifying implicit and explicit expressions of hope in all twenty-three speeches made by Israeli leaders in the Annual General Debate of the UN General Assembly. The second question is, where are the Palestinians regarding hope and skepticism toward peace?

The Politics of Hope and the Politics of Skepticism in Palestine

In 1974, Yasar Arafat was the first representative of a nongovernmental organization to speak at the UN assembly. In a later published manifesto about his UN speech, Arafat proclaimed he was fighting for freedom and hope. "I am a rebel, and freedom is my cause. . . . Why, therefore, should I not dream and hope?" (1975, p. 16). What was Arafat's hope? Arafat was clear that his dream for a peaceful future in Palestine meant liberating the land from Zionist imperialism and attaining political freedom for Palestinians. Arafat hoped for peace in the form of Palestinian independence though he assured that Jews would live peacefully in the State of Palestine. Years later, Abdel-Shai, one of the Palestinian delegates at the Madrid Peace Conference, reiterated Palestinians' hope for peace in the Holy Land: "We, the people of Palestine . . . have long harbored a yearning for peace and a dream of justice and freedom." Stated differently, the hope, or more accurately, the wish for peace, was the formal approach of the Palestinians in the early 1990s. Note that Abdel-Shai links peace with "justice" and "freedom." The connection between peace and structural issues such as justice and freedom is natural for low-power parties in conflict (Leshem & Halperin, 2020a). Indeed, for many Palestinians, peace is tantamount to justice, freedom, and self-determination.

During the heat of the Second Intifada, a joint Israeli-Palestinian track-two attempt called the *Geneva Initiative* tried to bring the idea of hope back into public discourse. The initiative began as a set of secret talks between Palestinian and Israeli public figures and former officials. Together they drafted a detailed blueprint for a new peace accord that addressed all critical issues in dispute. The accord was publicly launched in December 2003, in Geneva. Notably, the initiative sought to garner support for the idea of peace by reigniting the hope for it. "Peace is possible" was their motto.

Especially intriguing is the use of the rhetoric of hope among Palestinian leaders who were members of the Geneva Initiative or endorsed it after its launch. For example, Yasser Abed Rabbo (who, in the 1980s, led the Palestinian militant faction the Democratic Front for the Liberation of Palestine) has continuously spoken about reviving the hope (in this case, the expectations) for peace. "There is a chance to make peace," repeated Abed Rabbo in a series of video campaigns distributed online to the Israeli public by the Geneva Initiative. Other prominent Palestinian leaders conveyed similar messages of hope in this campaign. Interestingly, both dimensions of hope were articulated in their messages. For instance, Riyad al-Maliki, Foreign Minister of the Palestinian National Authority, stressed Palestinians' wishes for peace: "We are thinking about the future, not the past. . . . We want to proceed and achieve a lasting peace between our two peoples." Palestinians' expectations for peace were explicitly articulated by Saeb Erekat, who served as the Secretary-General of the Executive Committee of the Palestinian Liberation Organization (PLO). "I know that peace is doable. I know we can make it. . . . It can be done, and it will be done." Similar messages about the feasibility of peace were also made by Jibril Rajoub, former head of the Palestinian Preventive Security Force, and Dr. Sufian Abu Zaida, former Minister of Prisoner Affairs. At least in this campaign, which featured prominent figures from the Palestinian leadership, both dimensions of hope were front and center.

Palestinians' decision to promote hope for peace directly to the Israeli public was incentivized by their deep frustration with Israeli leadership, especially during Netanyahu's administration. Netanyahu's reluctance to promote peace led the PLO to establish a committee whose sole purpose was communicating directly to the Jewish Israeli public. Established in 2012, the Committee for Interaction with the Israeli Society (CIIS) was mandated to connect with Israeli media, civil society, and local leadership. The main message that the CIIS sought to convey is that the wish of the Palestinian leadership is to advance peace in the framework of the *Two-State Solution*. One of the committee's main rhetorical tools was the rhetoric of hope.

An excellent example of Palestinian leadership's politics of hope can be found in a 2019 interview with Muhamad Madani, the head of the CIIS. Speaking to the Israeli Press, Madani explains: "There are two trends in Palestine regarding future peace: hope and despair. The street is desperate, but the leadership is always optimistic, from Arafat to Abu-Mazen. At first, it was hope for [Israel's] eradication. Now it is the hope for partnership." Madani compared the Palestinian and Israeli approaches toward hope: "Contrary to the Palestinian hope, your [the Israeli] government spreads fear and pessimism. To advance their colonialist ideas, they are explicitly fighting every manifestation of joint hope. Their policy of promoting fear comes at the expense of faith and optimism." The analysis given by Madani, who is proficient in Hebrew and Israeli politics, suggests that the politics of skepticism promoted by Israeli leaders to their constituents was heard loud and clear in Palestine.

The attempt of a low-power party to communicate directly with society members of the high-power party deserves special attention. We have already mentioned that the main reason for this approach was the Palestinians' (probably correct) assessment that Israel's governments were reluctant to advance peace at the time. However, the bold act of expressing hope for peace directly to society members from the adversarial group needs to be understood from the perspective of the plight and suffering of the Palestinians. Between January 1, 2008 and January 1, 2022, more than 5,985 Palestinians lost their lives to the conflict (an average of 427 annually). This number is almost twenty times higher than the number of Israeli casualties in the same fourteen-year period. The number of Palestinians wounded in the same period is even more staggering: 132,840 (an average of 26 people daily). The high toll Palestinians are paying for the conflict and the hardships of living under military control suggest that Palestinians cannot simply abandon the hope for peace. They desperately need it.

Admittedly, expressions of hope for peace are highly contested among the Palestinian people and its leadership. As Muhamad Madani remarked, the Palestinian street is desperate. The disappointment from the Palestinian Authority and the never-ending experiences of oppression could easily make the most hopeful Palestinian talk about future peace with dismissal and cynicism. Moreover, Palestinian leadership is split, with Hamas ruling the Gaza Strip and its 1.8 million inhabitants for sixteen years (and counting). Hamas leaders have not spoken favorably about peace with Israel, let alone expressed hope for it.[11] However, for reasons described above, the politics of hope is still the main approach among Palestinian political elites.

We provided a brief overview of the politics of hope and the politics of skepticism in Israel and Palestine and offered our argument about the different approaches each side adopted regarding hope and skepticism. Building on this overview, we contend that, due to the moderate prices Israel pays for the conflict, Israeli leaders have been gradually adopting a skeptical stance toward peace. High-power parties who pay a relatively low price for the continuation of the conflict can afford to be skeptical. In many ways, the status quo serves the interests of the high-power groups (Kteily et al., 2013; Leshem & Halperin, 2020a), so a skeptical outlook regarding peace is conveniently in line with their interests. On the other hand, the Palestinians have been adopting the politics of hope because they cannot afford to be skeptical. Peace—and its assumed consequences: statehood and freedom from foreign control—is necessary for Palestinians. If the Palestinian cause for statehood is to be pursued, the politics of hope must be nurtured and endorsed.

To test these arguments, we sought to quantify the frequency of expressions of hope and skepticism toward peace made by Palestinian and Israeli leaders. As a first step, we searched for a corpus of speeches that can help us compare the frequency of expressions of hope and skepticism made by leaders from both sides. Our argument would be supported if Palestinians express wishes

[11] We thank Lior Lehrs for his insightful comment in this issue.

and expectations more than Israelis and refuted if the difference is in the other direction or if there is no difference between the two sides in the frequency of expressions of wishes and expectations for peace. We found this corpus in the archives of the United Nations (see Pressman, 2020, for a similar approach). Since 1998, representatives of the Israeli and Palestinian governments have been speaking about the conflict and regional politics in the annual debate of the UN General Assembly. Examining the frequency of hopeful expressions in the corpus would shed light on the political approach adopted by each side.

The Current Research

Each year, representatives of all member and non-member states in the UN receive fifteen minutes to address the UN General Assembly at its annual General Debate (Baturo et al., 2017). Israeli representatives have used the podium since 1949 to advocate for Israel and promote its standing in the international arena. Palestinians also used this stage to voice their plights and garner international support. However, being stateless, Palestinians were permitted to address the General Assembly only a few times.[12] In 1998, the Palestinians were invited to speak at the General Debates alongside the other Member States. Since then, delegates from both sides have used this highly public event to voice their perspective of the conflict and accuse their rivals of escalating violence and hampering peace. The UN's General Debate is the only venue where Israeli and Palestinian representatives can address, on equal footing, leaders of world countries. Relevant to our inquiry, speakers also used the opportunity to express their wishes and expectations for peace. This makes the speeches an excellent corpus for examining our hypothesis.

The texts of the speeches made at the annual General Debate comprise a suitable corpus for our purposes for several reasons. First, the speeches are always delivered over a few consecutive days, which means that representatives speak in a similar historical context each year. In addition, all speakers receive an identical amount of time to speak, and they all deliver their speeches to the same audience, which makes the comparison more robust. Last, representatives of both nations have spoken every year since 1998, meaning this is an exhaustive corpus with not one year missing.

We argued earlier that, at least to some extent, the past decades have been marked by the adoption of the skeptical approach by the Israeli leadership

[12] One exceptional occasion was Arafat's 1974 address to the UN, which was the first time a representative of an organization, in this case the PLO, was allowed the address the Assembly.

while the Palestinian leadership maintained the politics of hope. We believe these different approaches would be manifested in leaders' speeches in our corpus. More specifically, we hypothesize that, compared to Israeli speakers, Palestinian speakers will express their wishes and expectations for peace more frequently than their Israeli counterparts. However, it is essential to note that we did not anticipate that speakers from either side would openly say that they have no wish for peace. These types of statements are tantamount to diplomatic suicide. We also anticipated that speakers would mostly refrain from openly saying that peace is impossible, because announcing the demise of any future possibility of peace from the podium of the General Assembly might be diplomatically unwise. In other words, we did not anticipate many direct expressions of reluctance or pessimism for peace throughout the corpus.

However, we did anticipate that leaders from each nation would differ in the frequency of *affirmative* expressions of wishes and expectations for peace. Leadership that follows the politics of hope (postulated to be adopted by Palestinian representatives) will express more affirmative wishes and expectations for peace than leadership that follows the politics of skepticism (postulated to be adopted by Israeli representatives). Our hypothesis will be confirmed if Palestinian speakers do, in fact, utter affirmative expressions of hope more frequently than Israeli speakers.

Research Methods

In this study, we employed quantitative discourse analysis techniques (Fløttum, 2013; Johnson, 2011) to explore the explicit and implicit expressions of the two dimensions of hope.

Coding Procedure

The corpus includes forty-six speeches, twenty-three by each side, which are all the speeches delivered in the UN annual General Debates by Palestinian and Israeli representatives between 1998 and 2020. We used the formal UN translations to English from the UN archives. All speeches were coded by two independent coders trained for this task. Coders were instructed to detect expressions of hope for future peace. More specifically, coders identified speakers' expressions of the wishes for peace and the expectations for peace. Wishes and expectations for peace might be expressed explicitly or implicitly. Coders were thus guided to identify both explicit and implicit expressions of hope for peace in the texts. Levels of interrater reliability were above satisfactory (kappa = 0.715), and disagreements between coders were resolved by a third coder who was blind to the hypothesis.

Thought Units

Following Donohue and Druckman (2009), speeches were divided into *thought units*, which are the "minimum meaningful utterance having a beginning and an end" (Hatfield & Weider-Hatfield, 1978, p. 46). Usually, simple sentences were considered one thought unit, while more complex sentences were divided into several units. Overall, the corpus contained 5,866 units: 3,524 units by Israeli speakers, and 2,342 by Palestinian speakers. On average, speeches by Israeli representatives were about 1.5 longer ($M = 153$ thought units, $SD = 62$) than speeches by Palestinian representatives ($M = 102$ thought units, $SD = 48$). A t-test revealed that this difference is significant ($t = 3.13$, $df = 41.5$, $p = .003$). To increase comparability, we account for the different lengths of speeches in a later stage of analysis.

Coding of Context

Naturally, speakers talked about various issues, some unrelated to the Israeli-Palestinian conflict. We thus coded whether each thought unit was related to the conflict and omitted from further analysis those unrelated to the conflict.

Coding of Direction

Potentially, expressions of hope may come in an affirmative form (e.g., "We desire peace," "Peace is possible") or in a negative form (e.g., "We do not desire peace," "Peace is impossible"). Coders were guided to note whether hope was expressed in the affirmative or negative form. Table 6.1 provides examples of representatives' affirmative expressions of the two dimensions of hope.

Results

Descriptive Results

We first note that there was a significant difference between Palestinian and Israeli speakers in the number of thought units centered on the conflict. Palestinians devoted 76% of their speeches to the conflict (1,776 out of 2,342 thought units). In contrast, Israeli speakers dedicated only 33% of their speeches to the conflict (1,162 out of 3,524 thought units). A t-test comparing the mean number of thought units per speech devoted to the conflict shows

Table 6.1 Examples of units coded as expressing the two dimensions of hope

Hope dimension	Nationality of speaker	
Wish	Palestinian	In spite of all of this and despite the historical injustice that has been inflicted upon our people, their [Palestinians'] desire to achieve a just peace . . . will not diminish. *Abbas, 2010*
	Israeli	We want peace based on security and mutual recognition in which a demilitarized Palestinian state recognizes the Jewish State of Israel. *Netanyahu, 2013*
Expectations	Palestinian	We say before the international community that there is still a chance—maybe the last—to save the two-state solution and to salvage peace. *Abbas, 2012*
	Israeli	Practical progress toward peace is possible in those areas where there is an effective Palestinian Government that accepts the Quartet's principles. *Livni, 2007*

that this difference is significant (M_{PAL} = .74, *SD* = .15; M_{ISR} =.34, *SD* =. 23, *t* = 7.12, *p* <.01). This comparison alone suggests that the topic of the conflict is more pressing for Palestinians than for Israelis.

We now describe the frequency of expressions of hope for peace in the speeches. Across all speeches from both nationalities, hope for peace was expressed in 574 units. The wish for peace was expressed 331 times (182 times by Palestinian representatives and 149 times by Israeli representatives). Interestingly, the expressions of the wish for peace were always in the affirmative form, meaning that speakers never expressed their objection to peace from the podium of the UN General Assembly. Expectations for peace were expressed 243 times (137 by Palestinian representatives and 106 by Israeli representatives). As for expectations, representatives framed expectations in the affirmative form (e.g., peace is possible) 74% of the time, while the rest (26%) was in the negative form (e.g., peace is impossible).

Hypothesis Testing

Our primary hypothesis was that Israeli speakers would express less hope for peace than Palestinian speakers. To test our hypotheses, we first estimated an ANCOVA model where the speaker's nationality was entered as the independent variable and the overall expressions of hope (combining wishes and expectations) were entered as the dependent variable. To ensure that potential changes throughout the years would not confound the results, we controlled for the year in which the speech was delivered. In the second step, we sought to reveal a more nuanced understanding by comparing expressions of wishes for peace separately from expressions of expectations for peace.

Before estimating the models, we accounted for two factors. The first was the length of the speech. As mentioned, speeches delivered by Israeli representatives were on average longer than those delivered by Palestinians. To correct for the different lengths, we divided the number of hope expressions in a particular speech by the total number of thought units in said speech. In this way, we eliminate potential biases caused by the fact that Israeli representatives simply talked more. The second factor was the direction of the expressions of expectations for peace. Most expressions in our corpus appeared in their affirmative form (e.g., "Peace is possible"). However, there were also some expressions in the negative form when speakers expressed low expectations for peace. To correct the different directions, we subtracted the number of negative expressions of expectations for peace from the number of affirmative expressions of expectations.

Accounting for these two factors produced a score ranging between (−1) and (1), signifying the ratio of affirmative expressions of each variable per speech.[13] For example, a score of .03 in the expression of wishes for peace means that the speaker expressed the wish for peace in 3% of all thought units of a particular speech. The higher the score, the more pronounced the hope dimension in the speech. We used this score as the dependent variable in all analyses.

Results show that expressions of hope for peace (combing wishes and expectations) were more frequent in speeches made by Palestinian speakers than by Israeli speakers ($F\,(1, 22) = 4.887, p = .037$) (see Figure 6.1). Post hoc Tukey's HSD test found that Palestinians expressed their hopes for peace 1.75 times more than their Israeli counterparts ($M_{PAL} = .114$, $SD = .098$; $M_{ISR} = .065$, $SD = .044$, $p = .037$, 95% confidence interval [CI] = [.003, .095]). This means that, in complete support of our hypothesis, Palestinians expressed more hope for peace than Israelis in the twenty-three years the UN General Assembly speeches were made. The year the speech was delivered did not affect the frequency of expressions of the wish for peace ($F\,(22,22) = 1.03, p =. 465$).

We then ran similar models to uncover differences in the frequency of expressions of each hope dimension. Looking at the wish for peace, results show that, as hypothesized, expressions of wish for peace were more frequent among Palestinian speakers than Israeli speakers ($F\,(1, 22) = 7.977, p = .001$)

[13] Theoretically, the score can reach the values of (−1) or (+1). However, these values are unlikely since they mean that every sentence in the speech was about the hope (or lack of hope) for peace.

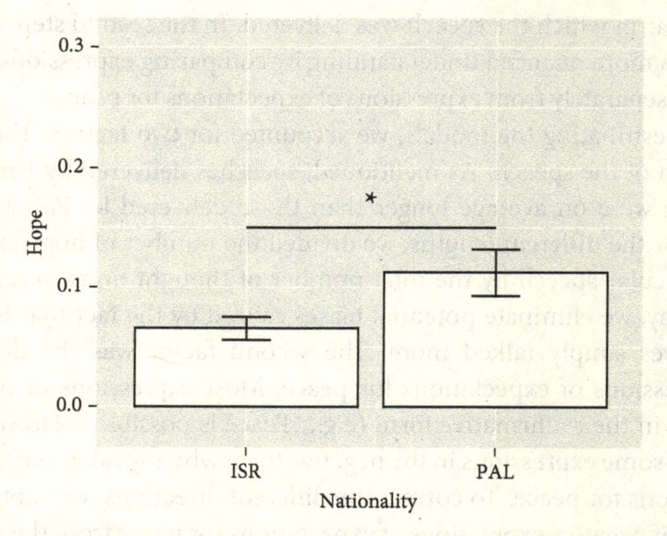

Figure 6.1 Average ratio of affirmative expressions of hope per speech by nationality of the speaker.

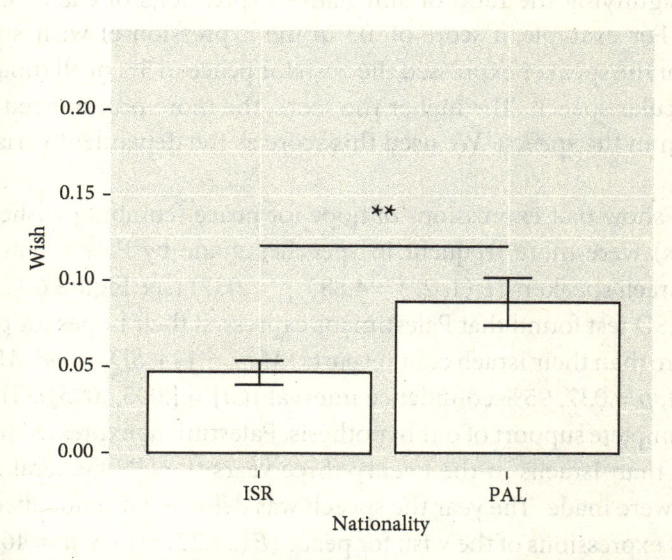

Figure 6.2 Average ratio of affirmative expressions of the wish for peace per speech by nationality of the speaker.

(see Figure 6.2). Post hoc Tukey's HSD test found that Palestinians expressed their wish for peace almost twice as much as Israelis (M_{PAL} = .087, SD = .065; M_{ISR} =. 046, SD = .033, p =.001, 95% CI = [.011, .071]). The year the speech was delivered was not associated with the frequency of expressions of wishes for peace (F (22,22) = 1.2, p =.335).

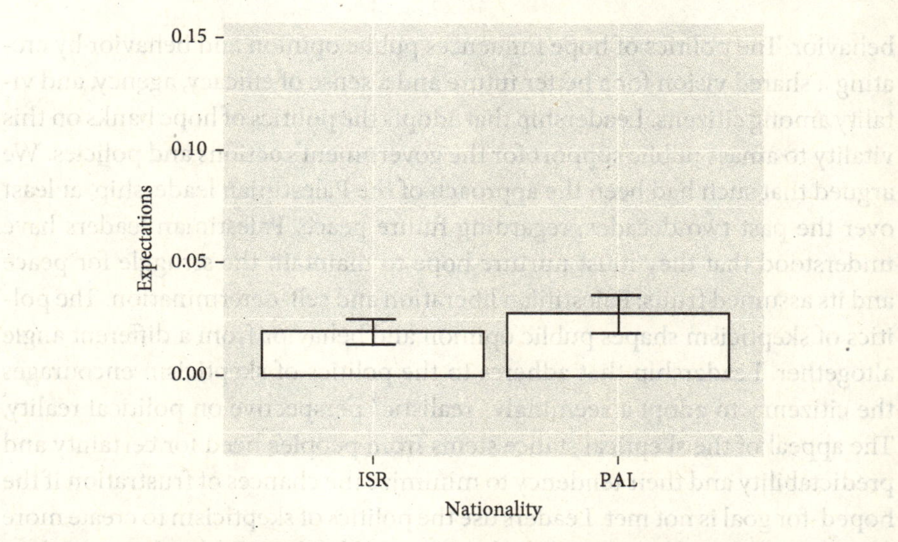

Figure 6.3 Average ratio of affirmative expressions of the expectation for peace per speech by nationality of the speaker.

When it comes to expressions of expectations for peace, results show that there was no significant difference in expressions of expectations for peace in speeches made by Palestinian and Israeli speakers (F (1, 22) = .69, p = .415) (see Figure 6.3). Palestinians' expectations for peace were more frequent, but the difference was not statistically different. The year the speech was delivered was not associated with the frequency of expressions of expectations for peace (F (22,22) = 1.069, p = .438).

Overall, consistent with our hypothesis, Palestinian speakers from 1998 onward expressed hope for peace more frequently than Israeli speakers. These findings provide initial evidence that the Palestinian leadership maintained the politics of hope while the Israeli leadership has deviated from hope toward a more skeptical approach. The postulated difference was observed in the overall expressions of hope and the expressions of the wish dimension of hope. This finding is revealing because it shows that, although representatives of both nations express a desire for peace, Palestinians express this desire more. There was no difference between Israeli and Palestinian speakers regarding speakers' expectations for peace. These points are discussed next.

Discussion

The politics of hope and the politics of skepticism are two approaches that may be consequential not only for governance style but also for public opinion and behavior. The politics of hope influences public opinion and behavior by creating a shared vision for a better future and a sense of efficacy, agency, and vitality among citizens. Leadership that adopts the politics of hope banks on this vitality to amass public support for the government's actions and policies. We argued that such had been the approach of the Palestinian leadership, at least over the past two decades, regarding future peace. Palestinian leaders have understood that they must nurture hope to maintain the struggle for peace and its assumed fruits: Palestinian liberation and self-determination. The politics of skepticism shapes public opinion and behavior from a different angle altogether. Leadership that adheres to the politics of skepticism encourages the citizenry to adopt a seemingly "realistic" perspective on political reality. The appeal of the skeptical stance stems from people's need for certainty and predictability and their tendency to minimize the chances of frustration if the hoped-for goal is not met. Leaders use the politics of skepticism to create more leeway to advance their agenda by lowering public demand for change and reform. We have contended that, in the past two decades, Israeli leaders have adopted the politics of skepticism toward future peace. As the high-power party in the conflict, Israelis readily adopted this skeptical stance. At least on the surface, Israel has little to lose by keeping the status quo.

The above arguments were tested in an analysis of the forty-six speeches made by Israeli and Palestinian representatives speaking at the General Debate of the UN General Assembly. Using quantitative discourse analysis, we compared the number of expressions of hope for peace made by Israeli and Palestinian speakers from 1998 to 2020. We first noted that when speakers talk about peace, both Israeli and Palestinian speakers expressed their desires for it and, most of the time, their belief in its possibility. This finding was not surprising, given that straightforward expressions of a skeptical stance might reflect negatively on the speaker and the entity she or he is representing. At least on the ceremonial occasion of the General Debate, delegates will try to avoid a formal expression of skepticism.

However, when we compared Palestinian and Israeli speakers, we found a significant difference in the frequency of their expressions of hope for peace. Combining expressions of wishes and expectations for peace, Palestinians' expressions of hope were 1.75 times more frequent than Israelis'. This finding supports our hypothesis and aligns with the statements Palestinian and Israeli leaders have made since the beginning of the millennia. Looking at each dimension of hope, it seems that this difference was driven mainly by the gap between Palestinian and Israeli speakers' expressions of the wish for peace, while differences in the expressions of expectations for peace did not reach statistical significance. In sum, statements about the possibility of peace were equally frequent among Palestinian and Israeli speakers, whereas statements about the desirability of peace were more common among Palestinian speakers.

Two points should be raised at this stage. The first is that, in practice, the politics of hope and the politics of skepticism often overlap. In the introduction, we presented the two political approaches as antithetical to simplify the distinction between them. However, leaders can adopt the two approaches simultaneously, even on the same subject (Botwinick, 2010; Oakeshott, 1996; Tseng, 2013). What is important is the big picture: namely, that an overarching pattern can be detected across time and changing governments. Our study shows that, across twenty-three years, Israelis and Palestinians expressed their hopes for peace at the UN General Debates, but Palestinians expressed hope more often.

The second point is that Israelis' and Palestinians' approaches toward future peace can be interpreted in several ways. The first is that the approaches reflect a genuine stance stemming from the power gap between the groups and the resulting difference in conflict-related experiences of each group. The stateless Palestinians, who live under harsh conditions of conflict, are likely to adopt the politics of hope because changing the status quo is essential for their existence. On the other hand, Israelis who experience the conflict in more moderate terms, fell less need to cling to hope. In both cases, the approaches can be regarded as genuine because they both derive from the different conflict-related realities experienced by Israelis and Palestinians. The second interpretation concerns the impact of the speeches on the audience back home. In the era of mass media, speeches given at the UN are transmitted live to domestic audiences in the speakers' home country. Leaders are, of course, aware of the media coverage their speeches receive and use the occasion to disseminate their agenda. The politics of hope and skepticism are thus distributed to Israeli and Palestinian publics by their respective leaders, furthering their hold on public opinion back home.

Last, when speeches are made to an international audience, they are also used to garner sympathy and support from the international community (Miskimmon et al., 2017; Nye, 2008). In these cases, the wording and rhetoric might be deliberately crafted to increase international attention and support for the conveyed narrative. This is likely to be pronounced when speakers represent states in conflict that naturally compete over international support and potential alliances (Faizullaev & Cornut, 2017; Friedman & Kampf, 2014; Nye, 2008). In this sense, hope for peace expressed by Palestinian and Israeli speakers should be understood as part of a "show," where speakers from each side would tend to portray their side as seekers of peace and conciliation. Our study demonstrates that this tendency is more pronounced among Palestinians, perhaps because they need international support more than do the Israelis. In other words, the asymmetrical nature of the conflict might affect the frequency of expressions of hope, even if these expressions are a part of a "show."

This chapter contributes to understanding hope amidst conflict in at least three ways. First, the theories of the politics of hope and the politics of skepticism add an important theoretical perspective to the discussion about hope in political contexts, including in the context of intractable conflicts. The theories point out that hope and skepticism should not be explored only as thoughts and feelings of individuals or collectives but also as deliberate leadership strategies. This perspective shines a unique light on the dynamics of conflicts and political processes more generally.

Second, by exploring hope and skepticism toward peace among leaders of parties in conflict, we add the layer of elite politics to the already discussed realms of hope in public opinion and grassroots activism. As political psychologists, we must explore political phenomena, including intractable violent conflicts, as bottom-up *and* top-down processes. In this regard, this chapter seeks to add a top-down perspective about hope amidst conflict to complement the bottom-up explorations in the book's other chapters. Third, the empirical investigation of hope in speeches by leaders of conflicting parties provides a novel way to study hope during conflict. So far, studies exploring hope in conflict have focused on public opinion and thus used surveys and experiments to reveal the role of hope amidst conflict (e.g., Cohen-Chen et al., 2014; Halperin et al., 2008; Leshem, 2019; Leshem & Halperin, 2020b). This chapter shed new light on how hope functions during conflict by analyzing transcripts of public speeches. .

Hopeful and skeptical comments made by public figures are often regarded as plain descriptions of political reality. When leaders or media pundits say that the chances for peace are low, people will tend to accept this interpretation as a fact-based comment made by those "who know." However, one of the premises of this book is that hope and hopelessness are also prescriptive; that is, they sanction and guide attitudes and behaviors. In this sense, public proclamation of skepticism toward peace made by Israeli leaders like Barak, Netanyahu, Bennett, Lapid, and others dangerously sways Israeli public opinion toward a skeptical, even cynical stance about future peace. The next chapter will reveal whether hope and skepticism affect people's policy preferences and political behavior during conflict.

References

Arafat, Y. (1975). The United Nations appeal for peace. In H. I. Hussaini (Ed.), *Toward peace in Palestine* (pp. 3–18). The Palestinian Information Office.

Baturo, A., Dasandi, N., & Mikhaylov, S. J. (2017). Understanding state preferences with text as data: Introducing the UN General Debate corpus. *Research & Politics*, *4*(2), 2053168017712821. https://doi.org/10.1177/2053168017712821

Bloch, E. (1959). *The principle of hope* (vol. 1–3). MIT Press.

Bloch, E. (1970). Man as possibility. In W. Capps (Ed.), *The future of hope* (pp. 50–67). Fortress.

Botwinick, A. (2010). *Michael Oakeshott's skepticism*. Princeton University Press.

Canovan, M. (1999). Trust the people! Populism and the two faces of democracy. *Political Studies*, *47*(1), 2–16. https://doi.org/10.1111/1467-9248.00184

Capps, W. (1968). The hope tendency. *Cross Currents*, *18*(3), 257–272.

Cohen-Chen, S., Halperin, E., Crisp, R. J., & Gross, J. J. (2014). Hope in the Middle East: Malleability beliefs, hope, and the willingness to compromise for peace. *Social Psychological and Personality Science*, *5*(1), 67–75. https://doi.org/10.1177/1948550613484499

Donohue, W. A., & Druckman, D. (2009). Message framing surrounding the Oslo I Accords. *Journal of Conflict Resolution*, *53*(1), 119–145. https://doi.org/10.1177/0022002708326443

Eckardt, M. H. (1982). The theme of hope in Erich Fromm's writing. *Contemporary Psychoanalysis*, *18*(1), 141–152. https://doi.org/10.1080/00107530.1982.10745682

Faizullaev, A., & Cornut, J. (2017). Narrative practice in international politics and diplomacy: The case of the Crimean crisis. *Journal of International Relations and Development*, *20*(3), 578–604. https://doi.org/10.1057/jird.2016.6

Fiske, S. (2010). *Social beings* (2nd ed.). Wiley.

Fløttum, K. (2013). *Speaking of Europe: Approaches to complexity in European political discourse*. John Benjamins.

Friedman, E., & Kampf, Z. (2014). Politically speaking at home and abroad: A typology of message gap strategies. *Discourse & Society*, *25*(6), 706–724. https://doi.org/10.1177/0957926514536836

Fromm, E. (1966). *You shall be as gods: A radical interpretation of the Old Testament and its tradition*. Open Road Media.

Fromm, E. (1968). *The revolution of hope*. Harper & Row.

Gamble, A. (Ed.). (2021a). Oakeshott and totalitarianism (2016). In *The Western ideology and other essays* (pp. 201–214). Bristol University Press. https://doi.org/10.46692/9781529217070.013

Gamble, A. (Ed.). (2021b). Oakeshott's ideological politics (2012). In *The Western ideology and other essays* (pp. 181–200). Bristol University Press. https://doi.org/10.46692/9781529217070.012

Goodman, M. (2018). *Catch-67: The left, the right, and the legacy of the Six-Day War*. Yale University Press.

Halperin, E., Bar-Tal, D., Nets-Zehngut, R., & Drori, E. (2008). Emotions in conflict: Correlates of fear and hope in the Israeli-Jewish society. *Peace and Conflict: Journal of Peace Psychology*, *14*(3), 233–258. https://doi.org/10.1080/10781910802229157

Hammer, D. C. (1995). The Oakeshottian president: George Bush and the politics of the present. *Presidential Studies Quarterly*, *25*(2), 301–313.

Hatfield, J. D., & Weider-Hatfield, D. (1978). The comparative utility of three types of behavioral units for interaction analysis. *Communication Monographs*, *45*(1), 44–50. https://doi.org/10.1080/03637757809375950

Johnson, K. (2011). *Quantitative methods in linguistics*. Wiley.

Kteily, N., Saguy, T., Sidanius, J., & Taylor, D. M. (2013). Negotiating power: Agenda ordering and the willingness to negotiate in asymmetric intergroup conflicts. *Journal of Personality and Social Psychology*, *105*(6), 978–995. https://doi.org/10.1037/a0034095

Laursen, J. C. (1992). *The politics of skepticism in the ancients, Montaigne, Hume, and Kant*. Brill. https://brill.com/view/title/2082

Leshem, O. A. (2019). The pivotal role of the enemy in inducing hope for peace. *Political Studies*, *67*(3), 693–711. https://doi.org/10.1177/0032321718797920

Leshem, O. A., & Halperin, E. (2020a). Lay theories of peace and their influence on policy preference during violent conflict. *Proceedings of the National Academy of Sciences*. https://doi.org/10.1073/pnas.2005928117

Leshem, O. A., & Halperin, E. (2020b). Hoping for peace during protracted conflict: Citizens' hope is based on inaccurate appraisals of their adversary's hope for peace. *Journal of Conflict Resolution*, *64*(7–8), 1390–1417. https://doi.org/10.1177/0022002719896406

Miskimmon, A., O'Loughlin, B., & Roselle, L. (2017). *Forging the world: Strategic narratives and international relations*. University of Michigan Press, p. 17.

Mullins, K., & Wildavsky, A. (1991). The procedural presidency of George Bush. *Society*, *28*(2), 49–59. https://doi.org/10.1007/BF02695502

Navot, D., Rubin, A., & Ghanem, A. (2017). The 2015 Israeli general election: The triumph of Jewish skepticism, the emergence of Arab faith. *The Middle East Journal*, *71*(2), 248–268. https://doi.org/10.3751/71.2.14

Nye, J. S. (2008). Public diplomacy and soft power. *Annals of the American Academy of Political and Social Science*, *616*(1), 94–109. https://doi.org/10.1177/0002716207311699

Oakeshott, M. (1996). *The politics of faith and the politics of scepticism. Oakeshott, M., & Fuller, T. (1996)*. (T. Fuller, Ed.). Yale University Press.

Oren, N. (2019). *Israel's national identity: The changing ethos of conflict*. Lynne Rienner.

Parsons, O. A., & Schneider, J. M. (1974). Locus of control in university students from Eastern and Western societies. *Journal of Consulting and Clinical Psychology*, *42*(3), 456–461. https://doi.org/10.1037/h0036677

Pressman, J. (2020). History in conflict: Israeli–Palestinian speeches at the United Nations, 1998–2016. *Mediterranean Politics*, *25*(4), 476–498. https://doi.org/10.1080/13629395.2019.1589936

Ravid, B. (26 October, 2015). Netanyahu: I Don't Want a Binational State, but We Need to Control All of the Territory for the Foreseeable Future. *Haaretz*. Retrieved May 23, 2023, from https://www.haaretz.com/israel-news/2015-10-26/ty-article/.premium/netanyahu-i-dont-want-a-binational-state-but-we-need-to-control-all-of-the-territory-for-the-foreseeable-future/0000017f-e67d-da9b-a1ff-ee7f93240000

Rorty, R. (1999). *Philosophy and social hope*. Penguin.

Rorty, R. (2002). Hope and the future. *Peace Review*, *14*(2), 149–155. https://doi.org/10.1080/10402650220140166

Rose, R. (1991). *The postmodern president: George Bush meets the world* (2nd ed). Chatham House.

Russell, B. (2001). Developing Derrida pointers to faith, hope and prayer. *Theology*, *104*(822), 403–411. https://doi.org/10.1177/0040571X0110400602

Senor, D., & Singer, S. (2011). *Start-up nation: The story of Israel's economic miracle*. Grand Central.

Shaw, T. (2007). *Nietzsche's political skepticism*. https://press.princeton.edu/books/paperback/9780691146539/nietzsches-political-skepticism

Tillich, P. (1965). *The right to hope*. Harvard Divinity School.

Tseng, R. (2013). Scepticism in politics: A dialogue between Michael Oakeshott and John Dunn. *History of Political Thought*, *34*(1), 143–170.

Zanany, O. (2018). *From conflict managing to political settlement managing: The Israeli security doctrine and the prospective Palestinian state*. Tami Steinmetz Center for Peace Research.

7
The Political Consequences of Hope

Everything that is done in this world is done by hope.

—Martin Luther

South African Apartheid was an established reality in 1966 when Robert F. Kennedy delivered his keynote address at the convention of the National Union of South African Students in Cape Town. Mandela was two years into his life-sentence imprisonment on Robben Island, and international pressure against South Africa was mounting. Unsurprisingly, Kennedy's visit to the county, at the request of the Student Union's president Ian Robertson, was severely frowned upon by South Africa's government. The Prime Minister was rightfully concerned that Kennedy's speech at a student conference would shine an unflattering light on the racist regime. On June 6, the young US senator stood in the crowded Jameson Hall of the University of Cape Town and delivered what most consider his most remarkable speech, also known as the Ripple of Hope speech.

> It is from numberless diverse acts of courage such as these that the belief that human history is thus shaped. Each time a man [sic] stands up for an ideal, or acts to improve the lot of others, or strikes out against injustice, he sends forth a tiny ripple of hope, and crossing each other from a million different centers of energy and daring those ripples build a current which can sweep down the mightiest walls of oppression and resistance.

Kennedy's Ripple of Hope speech conveys two notions. The first is that "small hopes" that might be powerless when isolated can amass into a tide of hope. The second is that this tide, this current of hope, is a potent political force that can effectively challenge oppressive social structures. Whether hope can aggregate is a question worthy of theoretical and empirical investigation. Nevertheless, the second claim about hope's potential to create change should deserve even more attention simply because it is hardly contested. People commonly take it for granted that hope can bring about positive outcomes.

Hope Amidst Conflict. Oded Adomi Leshem, Oxford University Press. © Oxford University Press 2024.
DOI: 10.1093/oso/9780197685303.003.0007

An implicit assumption that hope matters for social and political change resonated throughout this book's chapters.[1] Indeed, it makes intuitive sense that a desire for a political goal coupled with the belief in its possibility should affect people's political behaviors (e.g., their voting preferences) and positively impact political outcomes. However, given that the preceding chapters demonstrated that not all consequences of hope are positive, this assumption must be theoretically and empirically scrutinized.

Recall, for example, that many thinkers stressed that hope is desirable in its own right, regardless of whether it translates into broader desirable outcomes (e.g., Hume, 1740/2003; Mill, 1875; Tillich, 1965). In other words, hope should not only be judged by its objective consequences but also by the personal benefits it grants the hopeful person. Hope soothes and reassures and thus serves as a palliative in many stressful situations (Breznitz, 1986; Lazarus, 2013). If hope provides solace and can function as a coping tool to regulate anxiety and stress, people might settle for hoping for something rather than achieving it. In challenging political situations such as intractable conflict, people might even substitute their pursuit of peace for *hoping* for peace. There is a vast difference between the two. The former directs hope outward to create political change (see also van Zomeren, 2021). The latter directs hope inward to mitigate psychological discomfort. The interviews presented in Chapter 5 provide a flavor of this phenomenon.

Also recall that prominent thinkers highlighted the negative consequences of hope (Freud, 1927; Nietzsche, 1878; Spinoza, 1677). Hope might, for instance, send people chasing after unachievable goals instead of pursuing attainable ones. In the aftermath, when the unachievable goals are not met, disappointment and frustration might be insurmountable. If that is the case, it may be best to limit hope for peace in intractable conflict lest the setback is too harsh. Wishful thinking, the phenomenon in which people's expectations of achieving a goal are unduly exaggerated by their high wishes for the goal (Babad & Katz, 1991), is another undesirable consequence of hope. In this regard, hope might blind, distort assessments, deplete resources, and eventually create social stagnation. In sum, it should not be presupposed that hope has only positive consequences. This chapter's chief goal is to assess honestly the outcomes of hope.

More specifically, the main question posed in this chapter is: What, if any, are the positive and negative consequences of hope for peace? Most readers

[1] According to the bidimensional conceptualization of hope (see Chapter 3), hope has two dimensions: a wish to attain a goal and an expectation of attaining it. Consistent with the bidimensional conceptualization, I use the word "hope" to refer to the two dimensions at the same time. However, when I want to refer to only one dimension I do not use "hope" but the specific dimension: namely, "wishes" or "expectations."

might assume that hope has merit in the pursuit of resolution of conflicts, yet this assumption requires proof. There are several reasons why hope might not impact the advancement of peace. First, due to hope's palliative nature, hoping for peace might make the hopeful person feel less anxious and consequently leave the person politically passive. Specifically, hoping for peace might replace working for peace because hoping requires less engagement and investment. Second, it is also possible that peace-promoting outcomes are primarily influenced by factors other than hope and that hope has no or only a negligible impact on conflict resolution in the presence of more impactful factors. For instance, political ideology or religiosity, found to be robust predictors of conflict-related attitudes and behaviors (Leshem, 2017; Shulman et al., 2021), might overshadow the potential influence of hope. When the substantial influence of political ideology and religiosity are accounted for, the impact of hope might disappear. Finally, at least in theory, hope might exhaust mental resources and lead to political apathy. Under these circumstances, it may be wise to downplay the importance of hope in the overall pursuit of peace.

In the remainder of this chapter, I provide a detailed account of the potential consequences of hope by first citing research demonstrating that hope might have no (or negative) effects on political outcomes. I then outline several studies that show the contrary: namely, that hope has a substantial role in motivating political change. Next, I describe the findings from the Hope Map Project that examined the unique contribution of each of the two dimensions of hope in promoting attitudes conducive to peace. I then present experimental work that established a causal relationship between hope and peace-promoting attitudes, followed by a more detailed account of an experiment that showed how hope-inducing interventions translate into behaviors that facilitate peace and conciliation. I end by highlighting the gaps that still exist in studying the political consequences of hope.

Evidence for Null and Adverse Effects of Hope on Political Outcomes

The global struggle against the environmental collapse of the planet has caught the attention of the media in the past decade. Scholarly research on the different manifestations of environmental activism and its political, social, and economic underpinnings has also been increasing in number and scope (O'Brien et al., 2018; Ramelli et al., 2021). Relevant to the book's topic, several studies have explored the role of hope as a motivator of participation in environmental activism (e.g., Furlong & Vignoles, 2021; Kleres & Wettergren, 2017; Nairn, 2019; van Zomeren et al., 2019).

Overall, research on the psychological motivators of environmental activism shows that hope is not a significant predictor of people's engagement with environment-related action. For example, a study conducted in the United Kingdom among more than 200 people affiliated with the environmental Extinction Rebellion (XR) movement showed that hope was not related to activist intentions or behaviors (Furlong & Vignoles, 2021). In the study, agreement with statements such as "I feel hope that Extinction Rebellion will achieve its aims" was not associated with becoming active in the XR movement (or expressing an intention to do so). Similarly, across three experimental studies conducted in the United States, hope did not predict greater intentions to engage in environmental activism (van Zomeren et al., 2019). In these studies, participants read an expert's report claiming that climate change is reversible. Nevertheless, participants' "hope" (conceptualized here as the expectation that problems of climate change can be solved) was not associated with their willingness to participate in collective action and, in some cases, predicted less willingness to engage in environmental activism (see also Panagopoulos, 2014).

The palliative function of hope may explain the null or even negative association between hope and activism. As noted, hope is often used as an effective coping strategy to regulate stress (Breznitz, 1986; Lazarus, 2013). When hope functions as a coping mechanism, hope is directed inward to ease tension and anxiety. Perhaps believing that large-scale problems (be they climate change or intractable conflicts) can be solved alleviates some stress. When stress is alleviated, two interrelated processes might ensue. First, the perceived urgency of correcting the problem may simply decrease and, consequently, reduce one's willingness to take action. Second, the hopeful person might replace the taxing endeavor of pursuing a goal with the more pleasant feeling of hoping for it. Either way, hope does not always translate into action. This rationale led van Zomeran to conclude that hope might have a null effect on political activism in other domains (van Zomeren, 2021).

I agree that hope could have a null effect on political outcomes, mainly when hope is directed inward. When hope is used to soothe anxiety and make one "feel better," its potential to translate into political action is reduced. However, I have one central reservation regarding the null effect of hope. Studies that reported a null effect of hope on environmental activism conceptualized and operationalized hope as the belief in the possibility of stopping environmental disasters (Furlong & Vignoles, 2021; van Zomeren et al., 2019). Intentionally or unintentionally, measuring the extent to which one *wishes* to stop climate disasters was excluded from these studies. Yet, as the reader is now well aware, the level of hope to achieve a goal depends not only on expectations but is also intrinsically tied to the extent to which one wishes for it. It could thus definitely be the case that participating in environmental activism is linked to the wish dimension of hope and less to the expectation dimension. Because this link was not investigated, it is hard to know for sure.

In sum, findings from research on hope and environmental activism suggest that the expectation dimension of hope (i.e., believing that environmental problems are solvable) might translate into individual psychological healing at the expense of collective political action. Nonetheless, there is still no clarity about the link between the wish dimension of hope and collective action. Can the wish dimension have the same palliative effect? Could wishing for a political goal reduce psychological anxiety at the expense of pursuing the goal? As introduced in Chapter 3, these questions can only be answered by exploring hope as a bidimensional construct. Before turning to investigate the potential influence of each of the two dimensions of hope on political outcomes, I briefly describe studies in various domains that contradict the null effect of hope on political outcomes.

Evidence for Positive Effects of Hope on Political Outcomes

Notwithstanding the interesting findings demonstrating a lack of association between hope and political outcomes, a substantial body of research suggests otherwise (e.g., Courville & Piper, 2004; Greenaway et al., 2016; Halevy, 2017; Kleres & Wettergren, 2017; Moeschberger et al., 2005; Nairn, 2019; Wlodarczyk et al., 2017). For instance, a study conducted in Spain at the peak of the 15M demonstrations revealed the critical role of hope in driving collective action for social change (Wlodarczyk et al., 2017). Starting in 2011, the 15M movement rallied against government corruption and neo-liberal policies of the Spanish government (Castañeda, 2012). Data collected among more than 600 people, half of whom participated in the movement, show that hope was a robust predictor of participation in 15M demonstrations. The influential role of hope as a motivator for collective action is also supported by qualitative studies conducted in Sweden, Denmark, and New Zealand (Kleres & Wettergren, 2017; Nairn, 2019). Participants in these studies cite hope as a critical motivator for their participation in protests and activism.

Similarly, Greenaway et al. (2016) showed that hope is associated with support for political change in the United States and the Netherlands. In these studies, hope was a robust predictor of support for political change, even after accounting for other emotions. The researchers also used an experiment to show the discrete role of hope in eliciting support for social change. In the study, half of the participants were asked to write about something in their personal life that made them feel happy and half about something that made them feel hopeful. Participants in the "hope condition" were more willing to support actions and policies aimed at social change than those in the "happy condition" (Greenaway et al., 2016).

In the context of business interactions, Halevy (2017) demonstrated the positive effect of hope on reducing costly engagement in preemptive strikes. Utilizing an interactive decision-making task embedded in a series of experimental studies, Halevy shows that hope, but not anger or fear, reduced peoples' tendency to launch costly preemptive strikes against their business partners. Moving from business to conflict, Moeschberger et al. (2005) examined the association between hope and forgiveness among 300 Catholic and Protestant students living in Northern Ireland. Hope was found to predict Catholics' and Protestants' ability to decrease support for retaliation against the other side, which, in turn, predicted their willingness to forgive.

Empirical research demonstrating the positive link between hope and peoples' support for and engagement with social change corresponds with the premise that social evolution is hope-inspired (Bloch, 1959; Fromm, 1968). However, some research suggests that hope is unrelated to political change (van Zomeren, 2021; van Zomeren et al., 2019). Without undermining the importance of the studies discussed above, none of them explored hope as a bidimensional construct. Because of that, it is impossible to deduce from their results how hope, in its fullest form, relates to people's political attitudes and behaviors. As exemplified in previous chapters, the two dimensions are inseparable and both must be examined in any research on hope. The Hope Map Project conducted in conflict zones around the world utilizes the bidimensional model of hope to explore the predictive power of each dimension on conflict-related attitudes and behaviors. Because each dimension of hope is examined separately, we can accurately know whether and how conflict-related outcomes are associated with hope.

Hope as a Predictor of Peace-Promoting Positions

The Hope Map Project's goals, methods, and initial results were presented in Chapter 4. Here, I wish to provide a very brief recap. The Hope Map Project is an ongoing research project investigating hope for peace in conflict zones worldwide. The project seeks to identify the demographic and sociopsychological determinants of hope as well as its direct consequences on conflict-related attitudes and behaviors. The project's first phase was conducted in 2017, among representative samples of 500 Jewish Israelis and 500 Palestinians from the West Bank and the Gaza Strip. Its results are outlined in this book and elsewhere (Leshem, 2019; Leshem & Halperin, 2020a, 2020b, 2021). Data for the Hope Map Project are currently being collected among Turkish and Greek Cypriots, and planning has begun for projects in Colombia, India-Pakistan, and other conflict zones. The Hope Map Project offers the most comprehensive account of the hope of people mired in conflict.

Relevant to this chapter, the project examined the role of hope as a predictor of peace-promoting outcomes. I hypothesized that higher hopes for peace would predict Palestinians' and Israelis' peace-promoting attitudes in two domains. The first involved public support for concession-making as a part of the peace negotiation process. The second concerned the willingness to engage in peacebuilding projects. The following sections outline the methods used to test these hypotheses and the findings that emerged from the analyses.

Hope as a Predictor of Support for Compromise

Does hope for peace predict Jewish Israelis' and Palestinians' support for tough compromises that could lead to peace? In this study, I was interested in Israelis' and Palestinians' willingness to compromise on two highly contentious issues: the future of Jerusalem and the solution to the Palestinian refugee problem. Specifically, participants were asked how much they would support a compromise (on a scale from 1 to 5) on these issues if presented in a national referendum. The first item was "a peace agreement that would establish East Jerusalem as the Palestinian capital and West Jerusalem as the Israeli capital." The second item was about "a peace agreement that would include Israel's acknowledgment of its share in causing the refugee problem and accepting an agreed-upon number of refugees into its territory with compensation made to the remaining refugees from an international fund."

The central hypothesis was that the more people wish for peace and expect it to materialize, the more they would support compromises on these controversial topics. Importantly, I also tested whether the postulated association between hope and concession-making exists even after accounting for the influence of other factors like political ideology or religious observance, known to have a significant impact on the willingness to compromise (Leshem, 2017; Shulman et al., 2020, 2021). In the presence of such robust predictors, the postulated effect of hope on concession-making could disappear. If this is the case, hope's merit as a unique contributor to conflict resolution should be questioned.

Results

First, a gloomy picture emerges when looking at the overall means of support for compromise. Both Palestinians and Israelis were reluctant to compromise on Jerusalem and the refugees in return for peace. Palestinians' support for compromises was below the 3.0 midpoint of the scale ($M_{Jerusalem} = 1.89$, $M_{Refugees} = 2.54$), and so was Israelis' ($M_{Jerusalem} = 2.32$, $M_{Refugees} = 1.98$). Between-sample *t*-tests showed that Israelis were less supportive than Palestinians of the refugee compromise ($t = 6.44$, $p < .001$), whereas Palestinians were less supportive than Israelis of the compromise over Jerusalem ($t = 5.01$ $p < .001$). Interestingly, when the two items were collapsed into one variable indicating overall support for concession-making, Palestinians and Israelis seem to be equally reluctant to support compromise ($M_{Palestinians} = 2.22$, $M_{Israelis} = 2.15$, $t = 0.89$, $p = .37$).

The focus of our interest, however, was whether hope for peace predicts more willingness to compromise over these contentious issues. To test this question, a multivariate regression model was estimated for each national sample, where support for compromise served as the dependent variable; wishes and expectations for reciprocal peace served each as key predictors (see Chapter 4 for details); and political ideology, religious observance, political efficacy, acceptance of uncertainty, age, gender, and levels of education served as controls. Figures 7.1 and 7.2 present the standardized coefficient of variables that had a statistically significant effect on the support for compromise.[2]

As shown in Figure 7.1, the two dimensions of hope stand out as the strongest predictors of Palestinians' support for compromises on the future of Jerusalem and the Palestinian refugees (wishes: $\beta = .25$, $p < .001$; expectations: $\beta = .16$, $p < .001$). Public approval of compromises on these controversial

[2] Models explain 46% of variance in the Israeli sample and 17% in the Palestinian sample. Nonsignificant coefficients are not displayed.

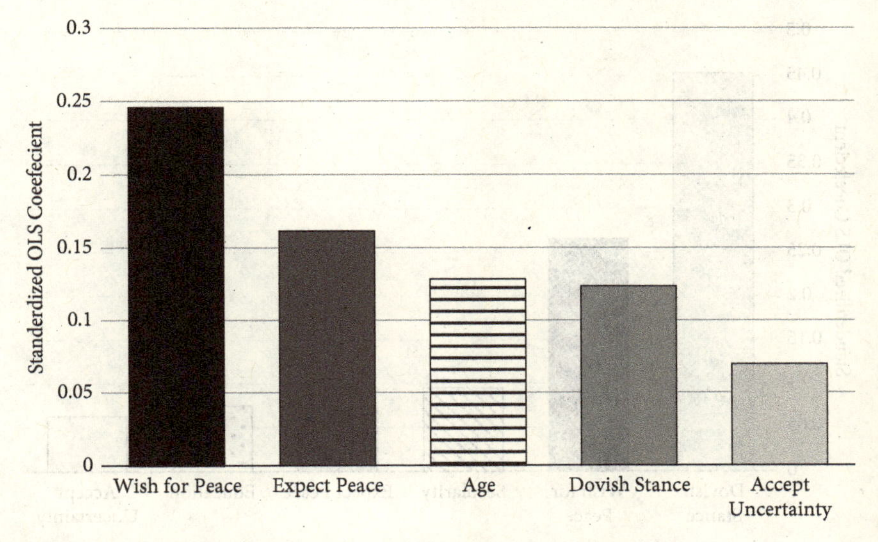

Figure 7.1 Predictors of Palestinians' support for compromise.

matters seems to be associated with the two dimensions of hope more than any other factor measured. Though Palestinians' age, dovish ideology, and acceptance of uncertainty were significant predictors of their willingness to make concessions, their impact was lower than the impact of hope for peace.[3] This finding offers initial evidence of the link between hope and conflict-related attitudes. Recall that existing research on the predictive power of hope explored only its expectation component, while the wish dimension was commonly overlooked. Findings presented here point to the substantial role of *both* dimensions in predicting conflict-related attitudes among people mired in protracted conflicts.

Next, looking at the Israeli sample (Figure 7.2), it seems that dovish ideological stance was the strongest predictor of support for compromise among Jewish Israelis ($\beta = .45, p < .001$). This strong association corresponds with existing research on the link between Israelis' dovish ideology and their support for a host of conflict-related policies (e.g., Pliskin et al., 2015; Shulman et al., 2021). However, even this strong relationship did not erase the effects of the wish ($\beta = .26, p < .001$) and the expectations ($\beta = .08, p = .035$) for peace on support for the referendum items. As can be seen, the effect of the wish dimension was much stronger than the effects of the expectation dimension (more than

[3] Age was a positive predictor of support for compromise (i.e., older participants tended to be more supportive of the referendum items ($\beta = .13, p < .01$)). Unsurprisingly, doves tended more than hawks to support compromise ($\beta = .12, p = .04$).

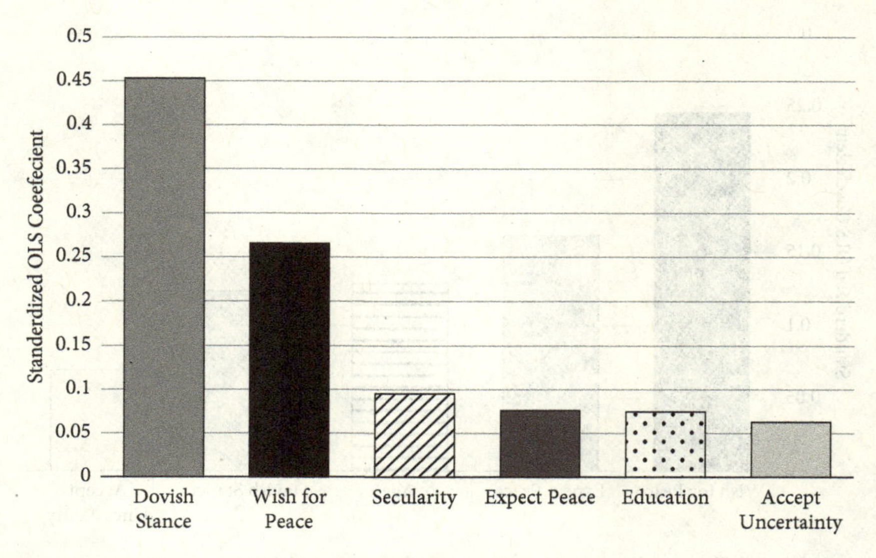

Figure 7.2 Predictors of Israelis' support for compromise.

three times in magnitude). Wald's test comparing the effects of each dimension reveals that this difference was statistically significant ($\chi2 = 8.03, p < .01$).[4]

Several initial insights can be drawn from looking at the findings so far. First, in line with previous studies (e.g., Canetti et al., 2015; Leshem & Halperin, 2020b), the data show that compromise is not popular among those living between the Jordan River and the Mediterranean Sea. This is not surprising given the mutual distrust between the two peoples (Oren, 2019; Shaked, 2018). The most revealing result is that the hope dimensions remain robust predictors of support for concession-making even after controlling for variables likely to predict concession-making. The wish dimension stands out as a substantial factor associated with people's willingness to forgo some of their most cherished principles while expecting peace to materialize contributes its share. Results provide initial evidence that supporting compromises is driven more by the desire to reach a reciprocal peace agreement than by the belief in reaching it. Support for compromise on pivotal issues is an essential part of promoting peace. Yet other aspects of conflict resolution, such as people's willingness to engage in peacebuilding projects, are also essential for the advancement of peace.

[4] It is also worth noting that Israeli seculars were more supportive of concession-making than the religious (β= .095, p = .011) and the educated more supportive than the less-educated (β= .074, p = .029). Similar to the Palestinian sample, Jewish Israelis who are generally open to uncertainty tend to support compromises more than those less comfortable with uncertainty (β= .063, p = .063).

Hope as a Predictor of Support for Peacebuilding

The second outcome of interest was Israelis' and Palestinians' support for peacebuilding initiatives. The question was whether hope's dimensions predict respondents' support for two initiatives. The first was a "joint non-violent demonstration calling for peace, justice, and security for all." The second was supporting "a team of Israeli and Palestinian ex-diplomats working on drafting a just and sustainable solution to the conflict." It was postulated that those with higher wishes and expectations for peace would be more supportive of these initiatives. Yet again, the critical point was to isolate the influence of the two dimensions from the influence of other variables, especially those likely to be highly correlated to Israelis' and Palestinians' engagement with peacebuilding projects. To test whether wishes and expectations for reciprocal peace predicted Israelis' and Palestinians' support for peacebuilding, I estimated a regression model identical to the previous one, with participants' support for peacebuilding initiatives serving as the dependent variable.

Results

Looking at the means, it appears that Palestinians and Jewish Israelis were generally unsupportive of the joint demonstration and the track-two diplomatic effort. Israeli's support was below the 3.0 midpoint of the scale for both initiatives ($M_{\text{demonstration}} = 1.97$, $M_{\text{track-two}} = 2.62$), and so was Palestinians' ($M_{\text{demonstration}} = 2.52$, $M_{\text{track-two}} = 2.42$). Between-sample t-tests showed that Israelis were less willing than Palestinians to support joint demonstrations ($t = 6.35$, $p < .001$) but that Palestinians were less supportive than Israelis of the track-two initiative ($t = 2.25$, $p = .02$). Collapsing the two items into a single variable indicating overall support for the peacebuilding initiatives shows that Palestinians were more supportive of peacebuilding compared to Israelis ($M_{\text{Palestinians}} = 2.47$, $M_{\text{Israelis}} = 2.29$, $t = 2.23$, $p = .026$). Moving to the main analysis, Figures 7.3 and 7.4 present predictors that had a statistically significant effect on participants' support for peacebuilding.[5]

Similar to the results for compromise support, the two dimensions of hope appear to be the strongest predictors of Palestinians' willingness to engage with peacebuilding projects (Figure 7.3). Again, the wish for reciprocal peace was the strongest predictor ($\beta = .25$, $p < .001$), and expecting reciprocal peace was the second strongest ($\beta = .18$, $p < .001$).[6] Turning to the predictors of support

[5] Models explain 35% of variance in the Israeli sample and 16% in the Palestinian sample. Nonsignificant coefficients are not displayed.

[6] Political efficacy predicted Palestinians' support for peacebuilding ($\beta = .15$, $p < .001$). This finding corresponds with studies showing the link between political efficacy and activism (van Zomeren, 2013) and is

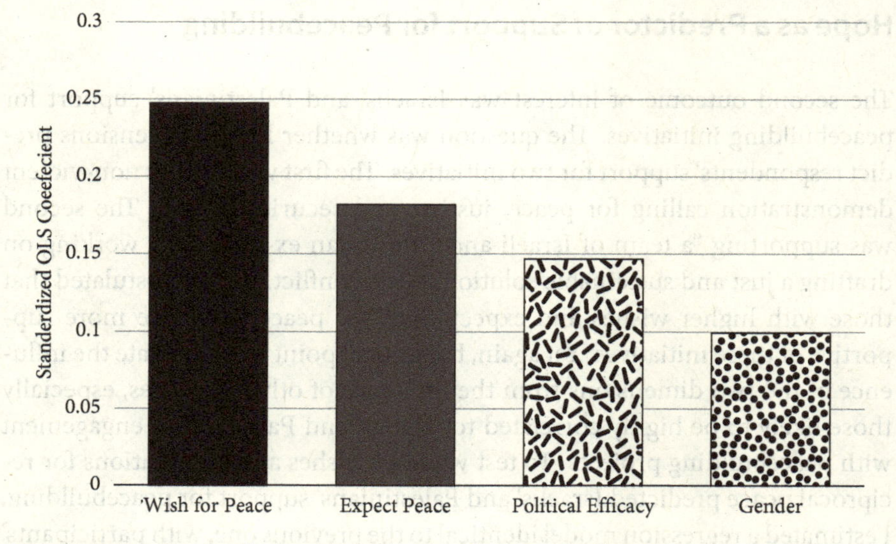

Figure 7.3 Predictors of Palestinians' support for peacebuilding.

for peacebuilding among Jewish Israelis (Figure 7.4), it appears that Jewish Israelis' desires and expectations for reciprocal peace predicted their support for the peacebuilding initiatives (wish: β= .28, $p < .001$; expectations: β= .18, $p < .001$) after accounting for the effects of all other factors. In other words, across variations in political stance and other potential predictors entered into the model, the two dimensions of hope for peace emerged again as essential contributors to public opinion on peacebuilding.

Summary of Results

The results suggest that hope is associated with advances in public support for peace in intractable conflict. In two rival populations, across two domains of peace-promoting outcomes, Palestinians' and Israelis' desires and expectations for peace emerged as robust predictors even after accounting for multiple factors such as religiosity, political ideology, and political efficacy. The hope components were among the most robust predictors of the peace-promoting outcomes, with the wish dimension standing out as a particularly central predictor. To obtain a broader look at the outcomes, I collapsed the

especially valuable considering the null effect of political efficacy on elite-level policies like compromises on Jerusalem and the refugee issues. Also, among Palestinians, men were more supportive of the peacebuilding initiatives compared to women (β= .1, $p = .019$).

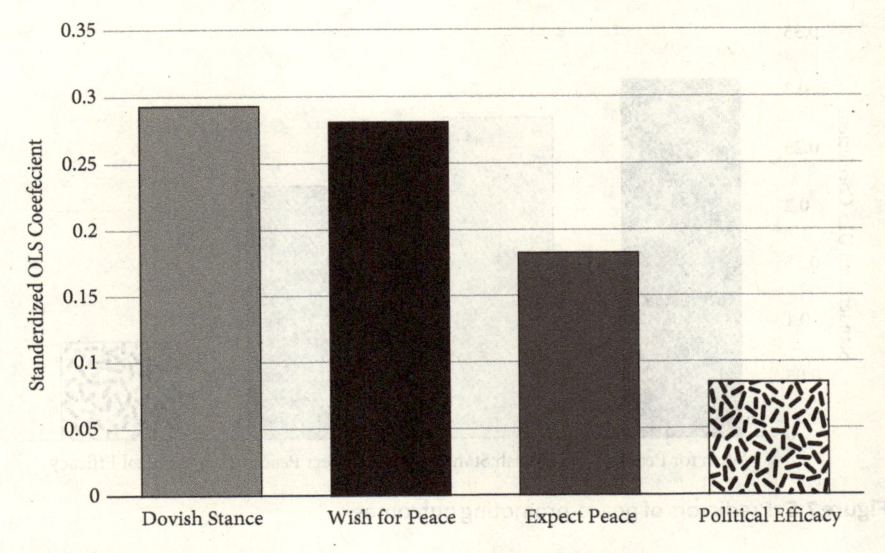

Figure 7.4 Predictors of Israelis' support for peacebuilding.

two peace-promoting domains (support for compromise and support for peace initiatives) into one dependent variable (α = .67) and re-estimated the regression model on the entire population (N = 1,000). Only four of the nine potential predictors emerged as significant (Figure 7.5, next page). Participants' wish for peace was the most robust (β= .31, p < .001), followed by dovish political stance (β= .28, p < .001), expectations for peace (β= .22, p < .001), and political efficacy (β= .08, p = .001).

Looking at all the figures presented in this chapter, we can cautiously say that the wish for peace was a stronger predictor of peace-promoting outcomes than the expectations for peace. Stating that the wish dimension is a more robust predictor of peace-promoting outcomes than the expectation dimension is a provocative statement that must be evaluated with caution because a statistically significant difference between the effects of each dimension was observed only in the case of Israeli's support for compromise (Figure 7.2). Still, analysis of unpublished data collected in 2021 among 600 Jewish Israelis shows that wishing for peace was a more robust predictor than expectations for peace on Israelis' support for a host of peace-promoting outcomes, such as giving humanitarian aid to the Palestinians, willingness to listen to the Palestinian narrative, and opposition to taking military action against Palestinians.

The difference between the wish and expectation dimensions as influencers of peace-promoting outcomes correspond to some degree with insights from

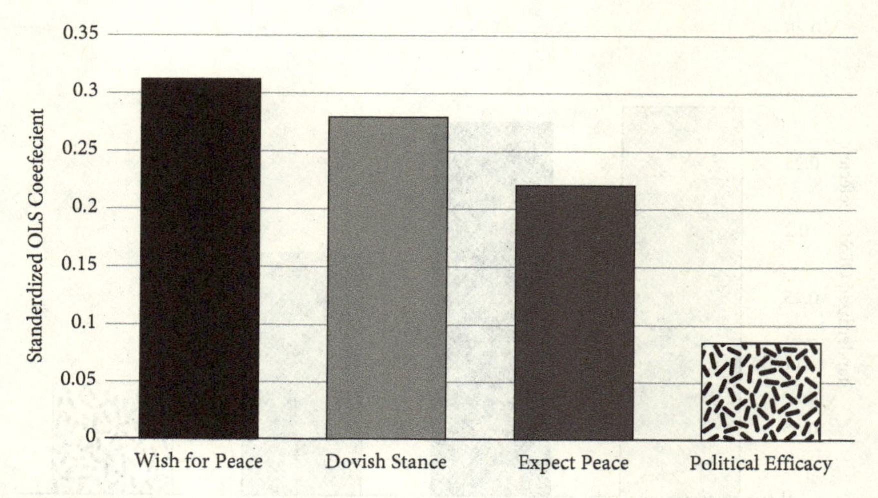

Figure 7.5 Predictors of peace-promoting outcomes.

the interviews with Palestinian and Israeli peace activists (Chapter 5). Among activists, the belief that peace could one day come was not the primary driver of their engagement in peace activism. Instead, their desires, aspirations, and eagerness for peace propelled them to be active. Overall, findings contribute to the critical debate about whether there is a hierarchy between the dimensions of hope. I expand on this point in the discussion.

Regardless of the relative predictive strength of the dimensions of hope, both emerged as robust predictors of attitudes conducive to conflict resolution. A direct theoretical and applied conclusion can be derived from the evidence linking hope and public support for peace-promoting outcomes, namely, that advancing peace could be achieved by advancing hope for it. This claim sounds logical, even intuitive. However, empirically demonstrating a causal link between hope and support of peace is more challenging than simply showing that the two are related. The following section describes experimental research conducted to establish causality between hope and public support for peace.

Hope for Peace as a Cause for Peace-Promoting Outcomes

The most common method used to establish causality is to experimentally manipulate (i.e., change) one variable (typically called the independent variable) and test the effects of such manipulation on another variable (the dependent variable). Thus, to establish causality between hope and support for peace in intractable conflicts, researchers need to raise participants' hope for peace and test whether this increase translates into positive changes in conflict-related attitudes and behaviors. This process requires two challenging steps. The first is to increase hope for peace. Although wishes for peace are relatively high in conflicts such as the one in Israel-Palestine, expectations for peace are meager. Convincing people immersed in a century-old conflict that peace is possible is a difficult task. Only if this challenge is surmounted is it possible to advance to the next challenging task, which is to test if this increase in hope (on one or both dimensions) translates into attitudes and behaviors that are more conducive to conflict resolution.

My first attempt to experimentally induce hope for peace in the context of the Israeli-Palestinian conflict was a complete failure. In a pilot study, seventy Jewish Israeli students read a fabricated newspaper article made to look like an article published in a popular Israeli news outlet. The article described a new study by Swiss Professor Susan Olaf, who explored the nature of 500 international conflicts over the past 300 years. According to the article, Olaf and her team from the University of Zurich succeeded in creating the most extensive comparative research on the characteristics of international conflicts. The Israeli reporter interviewing Olaf asked the professor what she could say about the characteristics of the Israeli-Palestinian case.

The participants in my experiment were unaware that they were randomly assigned to read one of two versions of Professor Olaf's answer. In one version, Professor Olaf revealed that the Israeli-Palestinian conflict fits the profile of international conflicts that were eventually resolved by a lasting peace agreement. In the second version, the professor revealed that the Israeli-Palestinian conflict fits the profile of international conflicts that were never resolved.

Results from the study show that although students rated the article's authenticity as high, both groups' expectations for peace remained similarly low. In other words, I have not succeeded in generating hope using reports of studies that claim that peace in Palestine-Israel is possible. There are at least two reasons why the article failed to induce hope (in this case, the expectation component). The first has to do with the communicator of hope. In this pilot experiment, the person communicating the information was a Swiss academic whom Israelis might consider as completely detached from what conflict reality is all about. A second reason is that the information conveyed in the manipulation was too straightforward. I tried to convince Israelis that peace is possible by seemingly "proving" it is. The message was perhaps too direct, too explicit, and thus more susceptible to provoking rejection. In hindsight, there is little surprise that it failed.

Studies that used indirect methods to induce hope for peace were more successful (Cohen-Chen et al., 2014, 2015). In these studies, Cohen-Chen and colleagues manipulated Jewish Israelis' belief in the malleability of conflicts and the world in general without mentioning the Israeli-Palestinian context. After participants read an article that conflicts, and the world in general, can change, their hope for peace in Israel-Palestine increased (though it is impossible to know which dimension of hope was affected). In another study, researchers induced hope for peace by asking people to draw a picture of the world as ever-changing and dynamic (Cohen-Chen et al., 2015). Notably, experimentally inducing hope increased Jewish Israelis' willingness to compromise on issues like borders and the future of Jerusalem (Cohen-Chen et al., 2014, 2015).

The above studies demonstrate that indirectly inducing hope for peace without mentioning the conflict or the adversarial group may increase support for compromise. The most significant advantage of indirect hope inducement is that it can reduce defensive reactions that might emerge when contentious topics are raised. Indeed, for group members immersed in prolonged conflict, mentioning "the conflict," "the Other," or the issues in dispute may evoke defensive mechanisms and rejection, subsequently hampering potential attitude change (de Zavala et al., 2008). It is therefore "safer" to induce hope for peace indirectly.

Nevertheless, real-world stimuli about the future of conflict and peace are always conflict- and group-specific so they naturally evoke defensive mechanisms. To advance conflict resolution theory and practice, it is thus essential to find ways to induce hope for peace without avoiding the contentious context of the conflict. Hence, a question arises whether hope for peace can be induced even when issues related to the conflict are mentioned and defensive mechanisms are activated. A study conducted immediately after my failed pilot experiment tackled this challenge more successfully.

Messengers of Hope, Conflict Escalation, and Peace-Promoting Behavior

My exploration into hope and hope inducement led me to design a hope-inducing intervention that does not avoid mentioning the conflict but instead uses the context of the conflict as a backdrop (Leshem et al., 2016). The study in question also addressed other limitations of existing research. First, past studies did not directly explore the effects that might be attributed to the *sources* of the hope-inducing cues. Stated differently, it was unclear whether effective hope inducement depends on the source communicating the hope-inducing information. Second, in all past studies, the measured outcomes of hope were attitudinal, not behavioral, so no evidence had been provided on whether increased hope affects political behavior. Third, hope was gauged immediately after participants were exposed to the hope-inducing interventions. Therefore, it is difficult to know if the effect of the interventions is long-lasting or transient. Designing hope-inducing interventions with sustainable effects over time is imperative in protracted conflict, where hope for peace must endure periods of hostilities and escalations.

The study was conducted in 2017 and addressed past limitations in three ways (Leshem, 2019). First, it examined whether the source of the hope-inducing information affects the effectiveness of hope inducement. More specifically, I sought to test who would be more effective in inducing hope for peace, a lay ingroup member or a lay outgroup member. Almost without exceptions, people favor messages conveyed by members of their own group (Nelson & Garst, 2005), whereas messages communicated by outgroup members, particularly disliked outgroups, are often rejected (Bohner & Dickel, 2011). If people are unlikely to be persuaded by disliked outgroup members, there is little reason to believe that Jewish Israelis will heed ideas conveyed by Palestinians. In fact, 79% of Jewish Israelis think Palestinians are untrustworthy and dishonest (Oren, 2019).

The logical conclusion is that messages aiming to induce hope for peace would be less effective when communicated by a member of the adversarial group than when conveyed by an ingroup member. However, as Deutsch et al. (2006) argued, a common fallacy held by people mired in conflict is that the enemy holds the keys to resolution. Based on this fallacy, escalation or de-escalation is assumed to lie entirely in the hands of the outgroup. It follows that when the feasibility of peace is discussed, opinions expressed by outgroup members will be perceived by ingroups as *epistemically superior* to opinions made by "one of their own." However disliked and distrusted, the outgroup might be regarded as a superior epistemic source when assessing the possibility of future peace. I thus hypothesized that an outgroup member would be more effective than an ingroup member in inducing hope for peace.

The second question was whether hope-inducing interventions could go beyond attitudinal change to elicit peace-promoting behavior. Most research shows that attitude change is easier to achieve than behavior change (Kruglanski et al., 2015). However, behavioral change is precisely what is needed to spur conflict transformation and peace. If attitudes are not translated into behaviors, no political change can transpire. Methodologically, measuring behavioral change usually requires a unique research design.

For this reason, I created a mock webpage that captured participants' active support for peacebuilding. The web page that "popped up" on participants' screens after completing the online survey contained a request from a Israeli-Palestinian peacebuilding initiative to support their social media campaign. Participants were made to believe their support for the campaign affected the chances of the peacebuilding initiative receiving funding from the United Nations. In this way, the measure of participants' support was located "outside" the survey to overcome a common problem in survey methodology where participants self-report their opinion for what they know are completely hypothetical questions (Fiske, 2010).

The third question concerned the durability of the effect of hope-inducing interventions. As noted, sustaining the effect of hope-inducing interventions beyond the immediate effect is vital in intractable conflicts where a lack of hope is the pervading assumption. Indeed, it is one thing to change attitudes and behaviors immediately after people are exposed to a hope-inducing message. It is much more challenging to sustain this influence when violence and hostility pull hope in the opposite direction. Initially, I sought to test the durability of the postulated effect of the hope-inducement messages in a follow-up study conducted a week after participants were exposed to the hope-inducing videos. However, an unexpected surge of violent confrontations in the West Bank and Israel proper, which began one day after participants were exposed to the hope-inducing messages, enabled the examination of a bolder question. Can hope-inducing interventions endure a week of conflict escalation and violent turmoil?

To test these questions, a representative sample of 426 Jewish Israelis participated in a study about their political opinions. As part of the study, respondents were asked to watch a two-minute video ostensibly uploaded to the internet by a video blogger expressing his view about the future of the conflict. For the study, I created four videos made to look like homemade blog videos. Two videos featured an Israeli blogger speaking in Hebrew, and two featured a Palestinian blogger conveying the same message in Arabic. Each blogger had two videos, one arguing that achieving peace is challenging but possible and the other that achieving peace is challenging and thus impossible.

Unknown to participants, they were randomly divided into five groups. Four groups were exposed to one of the four videos in a standard 2 × 2 experimental design. The design allowed me to test the effect of the message (peace is possible vs. peace is impossible), the messenger (an ingroup vs. an outgroup), and their interaction on participants' wishes and expectations for peace and active support for peacebuilding. The fifth group was not exposed to any video and thus served as a control group. Because the sample was representative of the Jewish Israeli population, the control group reflects Jewish

Israelis' baseline hope and support for peacebuilding. After watching the videos, respondents filled out a survey that included questions measuring their wishes and expectations for reciprocal peace. After a week of violent confrontations, the same panelists answered a seemingly unrelated survey about their wishes and expectations and were exposed to the mock "pop-up" webpage.

The surge of violence was sparked by an arson attack carried out by Jewish Israeli settlers in the Palestinian village of Duma. An eighteen-month-old Palestinian toddler died in the flames, while his four-year-old brother was severely wounded. The parents died some weeks later from their severe burns. Violent clashes between Palestinians and Israeli forces broke out when the news reported the atrocious attack. A Palestinian was shot and killed in Beir-Zeit after throwing a Molotov cocktail at an Israeli Defense Forces (IDF) post. The same day the IDF shot and killed a Palestinian and wounded another near the Gaza Strip. Rockets were fired from Gaza in retaliation. Demonstrations in East Jerusalem ensued, confronted by severe force from Israeli police. Some days later, a Molotov cocktail hit and injured a Jewish Israeli driver in Beit-Hanina. The Palestinian perpetrators declared that they sought to avenge the death of Ali Dawabsha, the eighteen-month-old victim of the Duma attack.[7]

When the follow-up survey ended, the pop-up webpage appeared on the participants' screen. The webpage advertised a social media campaign supporting an Israeli-Palestinian peacebuilding program. The campaign encouraged participants to help the peacebuilding initiative by asking them to allocate stars that appeared on the screen. The more stars allocated to the initiative, the higher the chances of the peacebuilding initiative winning a prized sponsorship from the UN. Participants could allocate any number of stars to support the peacebuilding initiative, decide not to allocate any stars, or close the pop-up webpage. The graphical appearance of the webpage was designed to make participants believe their decision would influence the success of a real-world Israeli-Palestinian peacebuilding initiative.

Results

For brevity's sake, I only report the findings from that part of the study collected after the violent escalation (for full results, see Leshem, 2019). Looking at the baseline group that was not exposed to any video ($N = 111$) reveals that Jewish Israelis' wishes for reciprocal peace were relatively high ($M = 4.54$,

[7] Escalation periods lasting several days to several weeks have become the norm in Israel-Palestine since the end of the Second Intifada. Some end with a small number of casualties, others end with thousands. Tragically, the fluctuation in the intensity of violence is one of the most constant and reliable feature of the conflict.

SD = 1.36 on a 1 to 6 scale), whereas expectations for peace were relatively low (*M* = 2.52, *SD* = 1.3 on the same scale). Jewish Israelis' active support of the peacebuilding campaign was rather low as well. On average, respondents allocated about two stars (*M* = 1.95; *SD* = 1.3) to the peacebuilding campaign out of a maximum of five.

Turning to examine the hypothesized effects of the video interventions shown before the escalation, it appears that when the communicator was Palestinian, the effects of the videos on participants' expectations for peace were significant ($b = 0.51$, $SE = 0.24$, $p = 0.04$) and marginally significant on their wishes for peace ($b = 0.43$, $SE = 0.23$, $p = 0.068$). This means that Israelis' expectations (and, to some degree, their wishes) for peace increased as a result of seeing an outgroup stating that peace is possible (vs. impossible). Recall that the participants were exposed to the short hope-inducing interventions a week earlier and that hostilities and violence in the region rose dramatically during that time. Thus, hope induced by the outgroup communicator withstood the escalation episode. Interestingly, the effects were not significant when the communicator was Israeli (expectations: $b = 0.15$, $SE = 0.22$, $p = 0.51$; wishes: $b = 0.29$ $SE = 0.27$, $p = 0.28$), indicating the ineffectiveness of an ingroup member to influence Israelis' wishes and expectations for peace.

Most notably, the hope-inducing interventions shown a week earlier affected Jewish Israelis' active support for peacebuilding. More specifically, an outgroup member's outlook on the future of the conflict had a significant effect on respondents' active support for the peacebuilding initiative ($b = 1.4$, $SE = 0.41$, $p < 0.001$), while the ingroup's outlook had no significant effect ($b = 0.35$, $SE = 0.42$, $p = 0.4$). In other words, after a week of hostilities, respondents' active support for peacebuilding increased due to the hope-inducing videos but only for those who saw the "enemy" saying that peace is possible. Moreover, those exposed to a Palestinian blogger saying that peace is possible exhibited significantly higher support for the peace initiative vis-à-vis the baseline group ($b = 0.85$, $SE = 0.14$, $p = 0.037$), while active support for peacebuilding among the three other treatment groups did not differ from baseline scores. In sum, the only effect observed was generated by the hopeful Palestinian blogger who increased Jewish Israelis' hope and consequently raised their active support for peacebuilding.

Discussion

The current study reveals the causal relationship between hope and peace-promoting behavior. Specifically, it shows that inducing hope for peace results in greater support of peace-promoting initiatives. It also highlights the pivotal role of the "enemy" in instilling hope for peace. In the videos, the Palestinian blogger identified as a Palestinian from Hebron, spoke Arabic, and did not express any particular sympathy toward Jews or Israelis. There is thus no reason to believe that participants had favorable opinions of him. However, it was the Palestinian, not the Israeli blogger, who induced hope for peace and elicited active support for peacebuilding. The "enemy" may be hated and distrusted (Oren, 2019), but the "enemy" seems to be highly effective when it comes to inducing hope for peace. This study also illustrates that lay opinions may affect the attitudes and behaviors of people mired in conflict. Although not tested vis-à-vis an authoritative figure, findings show that ordinary citizens can shape their adversary's beliefs and opinions. These findings align with research revealing the impact of social media on the attitudes of those living in prolonged conflicts (de Vries et al., 2017; Mor et al., 2016).

Perhaps most importantly, the study reveals that hope can have an impact beyond survey responses. Without exposure to a hope-inducing intervention, Jewish Israelis were quite reluctant to support what they believed to be a real-world peacebuilding project. However, after seeing hope-inducing messages conveyed by a Palestinian, their support increased significantly. Moreover, this increase was observed after a week of violent unrest, demonstrating the lasting effect of the hope-inducing interventions used in this study. Currently, the hope-inducing videos used in this study are being used as blueprints for peace-promoting campaigns in Israel-Palestine.

Notwithstanding the promising results of this study, some of its limitations should be mentioned. First, the dimensions of hope in the interventions were not completely isolated from each other. Although the bloggers spoke only about their expectations for peace, their wishes for peace might not have been effectively concealed. A second drawback is the relatively short period of escalation witnessed in the current study. Escalation episodes can last more than a week, so the current study can offer only limited inference on the durability of hope inducement. Third, the peace-promoting behavior used in this study gauged citizens' active support for a peacebuilding project, not their behaviors on more controversial issues. Most participants opposed the peacebuilding initiative (sadly indicating that even peoples-to-peoples programs are frowned upon by Jewish Israelis). However, more politically consequential behaviors must be explored to provide valuable knowledge on transforming intractable conflicts.

General Discussion

The first thing to say about the empirical study of hope's influence on conflict-related outcomes like public support for negotiation and peacebuilding is that the knowledge on the matter is still very limited. The studies outlined above indicate that hope matters, but to what extent and under what conditions is still unknown. It should be kept in mind that, at least in theory, hope can be directed at improving one's well-being at the expense of pursuing political transformation. Cohen-Chen et al. (2020) describe hope as a "feel-good" emotion that is also a "do-good" emotion. However, this feel-good emotion can also be directed inward, generating passivity and stalemate. Questioning the ability of hope to foster positive change corresponds with theoretical explorations of hope's dangers outlined in the writings of many thinkers (Fromm, 1968; Nietzsche, 1878; Spinoza, 1677). At the same time, most theoretical and empirical work shows that hope has direct positive consequences outside personal well-being (see Chapter 2 for a comprehensive review of the advantages and drawbacks of hope).

Findings from studies conducted among 1,000 Palestinians and Israelis in the framework of the Hope Map Project provide convincing evidence that hope is a robust predictor of peace-promoting outcomes. Respondents' desires for peace and, to a lesser extent, their expectations that peace will someday materialize were significantly associated with their support for compromise and peacebuilding steps. Experimental research further demonstrated hope's positive effects by revealing a causal link between hope and support for compromise and peacebuilding efforts (Cohen-Chen et al., 2014, 2015; Leshem, 2019). As such, hope can be regarded as a catalyst for positive outcomes in seemingly hopeless, violent conflicts. Nevertheless, the relevancy of the outcomes measured in the Hope Map Project and other studies on hope inducement can be questioned. Also questionable is the ability to generalize from experimental studies conducted only among Jewish Israelis to Palestinian society and other conflicts. In short, there is still much work to be done to identify the direct outcomes of hope amidst conflict.

Notwithstanding these limitations, I wish to highlight one insight I deem particularly valuable. Studies that differentiate between the two dimensions of hope provide evidence that the wish dimension is a stronger predictor of peace-promoting outcomes than the expectation dimension. In other words, when it comes to supporting compromise and peacebuilding, it is more important to wish for peace than expect it to materialize. The evidence is still limited and thus should be prudently evaluated. Nevertheless, the evidence points to a possible hierarchy between the dimensions.

The suggested hierarchy between the hope dimensions resonates with one of the main themes in the interviews with peace activists. Like many people locked in decades of intractable conflict, peace activists regarded the feasibility of peace as slim. On the other hand, activists' wishes and desires for peace were strong, vivid, and undeniable, and—as they expressed both implicitly and explicitly—were the main drivers of their activism. Vaclav Havel, an activist who became president of Czechoslovakia, concurs with the premise that wishing for a political goal is more important than expecting it to come. "Hope is not the conviction that something will turn out well," he said in an interview, "but the certainty that something makes sense, regardless of how it turns out" (Havel, 1990, p. 181).

Unfortunately, most research on hope in the context of conflict overlooked the wish dimension of hope, perhaps thinking that citizens' desires for peace are obvious or unvaried. However, as shown in Chapter 4, the desire for peace is neither obvious nor unvaried. This makes studying the consequences of *both* dimensions even more urgent. In the following, final chapter, I suggest several ways that the bidimensional model can be utilized in a host of political contexts.

References

Babad, E., & Katz, Y. (1991). Wishful thinking—Against all odds. *Journal of Applied Social Psychology, 21*(23), 1921–1938. https://doi.org/10.1111/j.1559-1816.1991.tb00514.x

Bloch, E. (1959). *The principle of hope* (vol. 1–3). MIT Press.

Bohner, G., & Dickel, N. (2011). Attitudes and attitude change. *Annual Review of Psychology, 62*(1), 391–417. https://doi.org/10.1146/annurev.psych.121208.131609

Breznitz, S. (1986). The effect of hope on coping with stress. In M. H. Appley (Ed.), *Dynamics of stress* (pp. 295–306). Plenum.

Canetti, D., Elad-Strenger, J., Lavi, I., Guy, D., & Bar-Tal, D. (2015). Exposure to violence, ethos of conflict, and support for compromise surveys in Israel, East Jerusalem, West Bank, and Gaza. *Journal of Conflict Resolution*, 0022002715569771. https://doi.org/10.1177/0022002715569771

Castañeda, E. (2012). The Indignados of Spain: A precedent to Occupy Wall Street. *Social Movement Studies, 11*(3–4), 309–319. https://doi.org/10.1080/14742837.2012.708830

Cohen-Chen, S., Crisp, R. J., & Halperin, E. (2015). Perceptions of a changing world induce hope and promote peace in intractable conflicts. *Personality and Social Psychology Bulletin, 41*(4), 498–512. https://doi.org/10.1177/0146167215573210

Cohen-Chen, S., Halperin, E., Crisp, R. J., & Gross, J. J. (2014). Hope in the Middle East: Malleability beliefs, hope, and the willingness to compromise for peace. *Social Psychological and Personality Science, 5*(1), 67–75. https://doi.org/10.1177/1948550613484499

Cohen-Chen, S., Pliskin, R., & Goldenberg, A. (2020). Feel good or do good? A valence-function framework for understanding emotions. *Current Directions in Psychological Science, 29*(4), 388–393. https://doi.org/10.1177/0963721420924770

Courville, S., & Piper, N. (2004). Harnessing hope through NGO activism. *Annals of the American Academy of Political and Social Science, 592*(1), 39–61. https://doi.org/10.1177/0002716203261940

de Vries, M., Kligler-Vilenchik, N., Alyan, E., Ma'oz, M., & Maoz, I. (2017). Digital contestation in protracted conflict: The online struggle over Al-Aqsa Mosque. *Communication Review*, *20*(3), 189–211. https://doi.org/10.1080/10714421.2017.1362814

de Zavala, A. G., Federico, C. M., Cisłak, A., & Sigger, J. (2008). Need for closure and competition in intergroup conflicts: Experimental evidence for the mitigating effect of accessible conflict-schemas. *European Journal of Social Psychology*, *38*(1), 84–105. https://doi.org/10.1002/ejsp.438

Deutsch, M., Coleman, P., & Marcus, E. (Eds.). (2006). *The handbook of conflict analysis and resolution* (2nd ed.). Jossey-Bass.

Fiske, S. (2010). *Social beings* (2nd ed.). Wiley.

Freud, S. (1927). *The future of an illusion*. Hogarth.

Fromm, E. (1968). *The revolution of hope*. Harper & Row.

Furlong, C., & Vignoles, V. L. (2021). Social identification in collective climate activism: Predicting participation in the environmental movement, Extinction Rebellion. *Identity*, *21*(1), 20–35. https://doi.org/10.1080/15283488.2020.1856664

Greenaway, K. H., Cichocka, A., van Veelen, R., Likki, T., & Branscombe, N. R. (2016). Feeling hopeful inspires support for social change: hope and social change. *Political Psychology*, *37*(1), 89–107. https://doi.org/10.1111/pops.12225

Halevy, N. (2017). Preemptive strikes: Fear, hope, and defensive aggression. *Journal of Personality and Social Psychology*, *112*(2), 224–237. https://doi.org/10.1037/pspi0000077

Havel, V. (1990). *Disturbing the peace: A conversation with Karel Hvížďala*. Vintage.

Hume, D. (1740/2003). *A treatise of human nature*. Courier.

Kleres, J., & Wettergren, Å. (2017). Fear, hope, anger, and guilt in climate activism. *Social Movement Studies*, *16*(5), 507–519. https://doi.org/10.1080/14742837.2017.1344546

Kruglanski, A. W., Jasko, K., Chernikova, M., Milyavsky, M., Babush, M., Baldner, C., & Pierro, A. (2015). The rocky road from attitudes to behaviors: Charting the goal systemic course of actions. *Psychological Review*, *122*(4), 598–620. https://doi.org/10.1037/a0039541

Lazarus, R. S. (2013). *Fifty years of the research and theory of R.s. Lazarus: An analysis of historical and perennial issues*. Psychology Press.

Leshem, O. A. (2017). What you wish for is not what you expect: Measuring hope for peace during intractable conflicts. *International Journal of Intercultural Relations*, *60*, 60–66. https://doi.org/10.1016/j.ijintrel.2017.06.005

Leshem, O. A. (2019). the pivotal role of the enemy in inducing hope for peace. *Political Studies*, *67*(3), 693–711. https://doi.org/10.1177/0032321718797920

Leshem, O. A., & Halperin, E. (2020a). Lay theories of peace and their influence on policy preference during violent conflict. *Proceedings of the National Academy of Sciences*. https://doi.org/10.1073/pnas.2005928117

Leshem, O. A., & Halperin, E. (2020b). Hoping for peace during protracted conflict: Citizens' hope is based on inaccurate appraisals of their adversary's hope for peace. *Journal of Conflict Resolution*, *64*(7–8), 1390–1417. https://doi.org/10.1177/0022002719896406

Leshem, O. A., & Halperin, E. (2021). Threatened by the worst but hoping for the best: Unraveling the relationship between threat, hope, and public opinion during conflict. *Political Behavior*, *20*. https://doi.org/10.1007/s11109-021-09729-3

Leshem, O. A., Klar, Y., & Flores, T. E. (2016). Instilling hope for peace during intractable conflicts. *Social Psychological and Personality Science*, *7*(4), 303–311. https://doi.org/10.1177/1948550615626776

Mill, J. S. (1875). *Theism* (R. Taylor, Ed.).

Moeschberger, S. L., Dixon, D. N., Niens, U., & Cairns, E. (2005). Forgiveness in Northern Ireland: A model for peace in the midst of the "Troubles." *Peace and Conflict: Journal of Peace Psychology*, *11*(2), 199–214.

Mor, Y., Ron, Y., & Maoz, I. (2016). "Likes" for peace: Can Facebook promote dialogue in the Israeli–Palestinian conflict? *Media and Communication*, *4*(1), 15. https://doi.org/10.17645/mac.v4i1.298

Nairn, K. (2019). Learning from young people engaged in climate activism: The potential of collectivizing despair and hope. *Young*, *27*(5), 435–450. https://doi.org/10.1177/1103308818817603

Nelson, T. E., & Garst, J. (2005). Values-based political messages and persuasion: Relationships among speaker, recipient, and evoked values. *Political Psychology*, *26*(4), 489–516.

Nietzsche, F. (1878). *Human all too human*. Gordon.

O'Brien, K., Selboe, E., & Hayward, B. M. (2018). Exploring youth activism on climate change: Dutiful, disruptive, and dangerous dissent. *Ecology and Society*, *23*(3). https://www.jstor.org/stable/26799169

Oren, N. (2019). *Israel's national identity: The changing ethos of conflict*. Lynne Rienner.

Panagopoulos, C. (2014). Raising hope: Hope inducement and voter turnout. *Basic and Applied Social Psychology*, *36*(6), 494–501. https://doi.org/10.1080/01973533.2014.958228

Pliskin, R., Sheppes, G., & Halperin, E. (2015). Running for your life, in context: Are rightists always less likely to consider fleeing their country when fearing future events? *Journal of Experimental Social Psychology*, *59*, 90–95. https://doi.org/10.1016/j.jesp.2015.04.001

Ramelli, S., Ossola, E., & Rancan, M. (2021). Stock price effects of climate activism: Evidence from the first Global Climate Strike. *Journal of Corporate Finance*, *69*, 102018. https://doi.org/10.1016/j.jcorpfin.2021.102018

Shaked, R. (2018). *Behind the kaffiyeh: The conflict from the Palestinian perspective*. Yediʿot aḥaronot: Sifre ḥemed.

Shulman, D., Halperin, E., Elron, Z., & Reifen Tagar, M. (2021). Moral elevation increases support for humanitarian policies, but not political concessions, in intractable conflict. *Journal of Experimental Social Psychology*, *94*, 104113. https://doi.org/10.1016/j.jesp.2021.104113

Shulman, D., Halperin, E., Kessler, T., Schori-Eyal, N., & Reifen Tagar, M. (2020). Exposure to analogous harmdoing increases acknowledgment of ingroup transgressions in intergroup conflicts. *Personality and Social Psychology Bulletin*, *46*(12), 1649–1664. https://doi.org/10.1177/0146167220908727

Spinoza, B. (1677). *Ethics* (G. H. Parkingson, Trans.). Oxford University Press.

Tillich, P. (1965). *The right to hope*. Harvard Divinity School.

van Zomeren, M. (2013). Four core social-psychological motivations to undertake collective action. *Social and Personality Psychology Compass*, *7*(6), 378–388. https://doi.org/10.1111/spc3.12031

van Zomeren, M. (2021). Toward an integrative perspective on distinct positive emotions for political action: Analyzing, comparing, evaluating, and synthesizing three theoretical perspectives. *Political Psychology*, *42*(S1), 173–194. https://doi.org/10.1111/pops.12795

van Zomeren, M., Pauls, I. L., & Cohen-Chen, S. (2019). Is hope good for motivating collective action in the context of climate change? Differentiating hope's emotion- and problem-focused coping functions. *Global Environmental Change*, *58*, 101915. https://doi.org/10.1016/j.gloenvcha.2019.04.003

Wlodarczyk, A., Basabe, N., Páez, D., & Zumeta, L. (2017). Hope and anger as mediators between collective action frames and participation in collective mobilization: The case of 15-M. *Journal of Social and Political Psychology*, *5*(1), 200–223. https://doi.org/10.5964/jspp.v5i1.471

8
Conclusion

> The more unpropitious the situation in which we demonstrate hope, the deeper that hope is.
>
> —Vaclav Havel

I began writing this book in May 2021, while Israeli Defense Force planes were bombing the Gaza Strip and missiles from the Gaza Strip were being fired at Israeli towns. My family and I were in and out of the bomb shelter for two weeks. The routine was simple. Once the sirens' insistent cry was heard, we had ninety seconds to reach the shelter. That left me enough time to grab the kids and my laptop. And so some of my critical thoughts about hope for peace were written to the sound of missile explosions clearly audible from the well-protected bomb shelter in our house in Tel Aviv. On May 21, when the ceasefire went into effect, Palestinians and Israelis counted their dead. More than 230 were killed in Gaza and 12 in Israel.

A relatively calm period followed in Israel and the Occupied Palestinian Territory (OPT), which naturally created some sense of positivity and hope. I admit it was easier to write about hope when the drums of war were not banging as hard, perhaps because the contradiction between the bloody reality of conflict and the hoped-for peace was less apparent. The chapters were coming along, and the book was slowly taking shape. Nothing prepared me for what happened at 9 PM on April 7, 2022, when a twenty-eight-year-old Palestinian from the Jenin Refugee Camp went on a shooting spree 500 meters from our home. The shooter killed three civilians, fled, and was believed to be hiding in the backstreets of our neighborhood. Hundreds of police and military officers stormed the streets with locked and loaded guns, helicopters hovered over our house, and a loudspeaker ordered everyone to stay inside and lock all windows and doors. It was frightening for all of us, but for me, it was also very depressing. How can advocates of peace advance resolution when tensions and hostilities are so high? How can I write about hope when blood flows in the streets of Jenin and Tel Aviv?

Hope Amidst Conflict. Oded Adomi Leshem, Oxford University Press. © Oxford University Press 2024.
DOI: 10.1093/oso/9780197685303.003.0008

In the morning, they found and shot him. Several days later, an Israeli general issued an order to demolish the perpetrator's house. Demolishing houses belonging to the families of Palestinian perpetrators is a brutal and impractical practice. However, it is a standard procedure implemented by Israel after lethal attacks such as the one carried out near our home. The formal reason for these collateral punishments is deterrence, yet house demolitions have been shown to increase, not diminish Palestinian aggression (Hatz, 2019). Not surprisingly, the tensions in the West Bank grew as a consequence, as did the number of Palestinian casualties. Four weeks later, two Palestinians from the Jenin area entered Israel proper and killed three Jewish civilians with knives and axes.

Is it legitimate to talk about hope for peace while people are burying their loved ones? Is hope a thing to be discussed when hope is essentially absent? If so, should hope be axiomatically considered beneficial (as is often the case) or something that should be avoided to enable the day-to-day endurance of those enmeshed in conflict? Perhaps hope does not matter at all. During the writing process, I asked myself these questions, questions I believe are unavoidable for any scholar and practitioner working on conflict and peace. Good scholarship and effective practice must always be connected to harsh reality, not some sugar-coated version of it. In this book, I tried to generate an honest discussion about the link between the cruel reality of intractable conflicts and the wishes and expectations to transform them.

The fundamental questions about hope's merits, dangers, and contribution to conflict transformation arise not only in the case of Israel-Palestine but also in other contemporary international conflicts, such as the evolving war in Ukraine or the frozen and intractable dispute in Cyprus (Heraclides, 2011). Fundamental questions about hope are first ruminated in the minds of those most affected by the conflict, namely, citizens and leaders living amidst conflict and violence. Fundamental questions about hope are also asked by third-party actors and international stakeholders who wonder how the conflict will unfold and whether they will profit from its resolution. Notably, these questions are asked on both dimensions.

The first question pertains to the extent to which peace is desired. Here, leaders, citizens, and international actors are asking: How desirable is peace? This book reveals that strong desires for peace, among both the masses and elites, cannot be assumed. Results presented in this book reveal that some people have no desire for peace, even when the definition of peace is entirely open-ended (Leshem, 2017). Moreover, looking at the wish for peace through the lens of asymmetrical power relations, it appears that strong wishes for peace are likely to be expressed more by disadvantaged parties and their allies because the urgency for resolution is acute for those who pay the highest price in the conflict. The advantaged, on the other hand, are inclined to be more moderate in their desires for peace because, from their point of view, the price of conflict is bearable and, in extreme asymmetric cases, even negligible.

The question of hope is also asked on the expectation dimension of hope. When it comes to expectations, citizens, leaders, and other stakeholders are asking to what extent peace is possible. They evaluate the chances for peace and contemplate the conditions that might make peace feasible. However, the answers they give are not only descriptive but, to a large extent, very prescriptive. If peace is described as something feasible, people might try to bring it about. If it is deemed impossible, no one would even try.

Rethinking hope is, therefore, crucial for understanding conflicts and the possible ways to transform them. For at least two reasons, revisiting the topic of hope is also essential for basic social science theory and empirical research. In terms of theory, problematizing hope was needed because, almost without exception, hope is a priori assumed to be a positive thing, something humans should always cherish and strive for, no matter the cost. However compelling, this view is partial and culturally presupposed and so deserves further scrutiny if we seek to fully understand hope and its boundaries. In this book, I have taken the task of de-romanticizing hope. In doing so, I believe hope's advantages (and pitfalls) emerged more clearly. The book's chapters reveal that, alongside its merits and unique power, hope has limitations and blind spots that should be acknowledged and accounted for in broader theoretical investigations.

In terms of empirical research, reexamining the concept of hope was needed to reveal some inconsistencies in the existing empirical investigation of hope. Empirical investigation requires measuring tools. When these tools are inaccurate, results and their interpretations can mislead. The recent increase in scholarly attention to the question of hope during conflict made it even more necessary to offer a comprehensive and concise conceptualization that can be used in the empirical study of hope. One of the ways this book contributes to social science is by providing the bidimensional model of hope and presenting its utility in quantitative and qualitative research.

The remainder of the chapter begins with a short recap of the book's chapters and continues with answering the overarching research questions posed in the Introduction. I then expand on the book's broader theoretical and empirical contributions to political psychology and conflict analysis and suggest ways that insights from this book can help practitioners working in the field. I end with some thoughts about *optimal hope* and how it could be applied in scholarship and practice.

Recap

The introductory chapter set the stage for our journey into the nuanced relationship between hope and conflict. It also articulated the main research questions that drive this book's theoretical and empirical inquiry. Next, Chapter 2 reviewed the benefits and downsides of hope as explicated by philosophers and political thinkers and as exemplified in situations of conflict. The merits of hope are indeed more intuitive and, at least in the West, are perceived as self-explanatory. The dangers of hope, including ignorance (Descartes, 1649/2015, Spinoza, 1677/2000), arrogance (Gravlee, 2020), and anguish (Nietzsche, 1878/1974), are less obvious, and so a particular emphasis was given to hope's drawbacks.

Given that these benefits and shortcomings are intrinsic to hope, the question that arises is how the advantages could be amplified and the pitfalls avoided. Answering this question becomes crucial when hope is fragile, such as during prolonged conflicts. In conflict situations, the question becomes "How can hope be sustained and impactful?" One solution to manage hope in the challenging circumstances of conflict comes in the form of *optimal hope.* In Chapter 2, I began to explain what I mean by optimal hope. I continue to expand on the matter at the end of this chapter.

Hope's properties and structure were scrutinized in Chapter 3. Using psychological and philosophical inquiry, I outlined how hope has been conceptualized in existing literature and showed the problems with these conceptualizations, especially in existing research on hope in intergroup conflict. Distilling the elemental components discussed by philosophers and psychologists, I offered a parsimonious but comprehensive conceptualization of hope in the form of the bidimensional model, best represented on a bidimensional plane. I provided the rationale for the bidimensional model of hope and described its implementation as a measuring tool. The chapter ended by demonstrating the model's utility in comparing the hopes for peace among two rival societies locked in an intractable dispute: Israelis and Palestinians.

Chapter 4 focused on a fundamental question: "What are the determinants of hope?" Utilizing data from the Hope Map Project, the chapter revealed some demographic and sociopolitical predictors of hope for peace, such as age, religiosity, and political ideology. Drawing on the bidimensional model of hope, the chapter also demonstrated that factors like acceptance of uncertainty, political efficacy, and perceptions of threat correlate with one dimension of hope but not with the other. Several explanations were provided for the nuanced relationship between the hope dimensions and their correlates. The chapter ended with a discussion about the finding that low-power group members, in this case Palestinians, have higher expectations for peace than the high-power Israelis. A deeper look into the relationship between power and hope is offered in the subsequent sections of this chapter.

Hope's role in grassroots peacebuilding is the topic of Chapter 5. Written in collaboration with Shanny Talmor and Eran Halperin, the chapter explored hope from the perspectives of activists working on the frontline of peacebuilding. The thematic analysis of the twenty interviews with Palestinian and Israeli peacebuilders told an eye-opening story about how hope is used and interpreted by peace activists. The findings exposed the palliative aspects of hope and the reciprocal relations between hope and political activism. Hope drives activists' commitment to action. Their actions, in turn, replenish their hopes. Integrating the findings, we raised a provocative thought: maybe hope, not necessarily political transformation, is the end target of activism.

Chapter 6 took the reader from the unadorned offices of peacebuilding nongovernmental organizations (NGOs) to the elegant hall of the United Nations General Assembly in New York. Written together with Ilana Ushomirsky, Emma Paul, and Eran Halperin, the chapter examined how hope and skepticism are used by leaders of nations in conflict. We first expanded on the politics of hope and the politics of skepticism as strategies used by political elites. We then provided a brief historical account of how Israeli and Palestinian leaders implemented the two strategies. Next, we compared the frequency of hope expressions made by Israeli and Palestinian delegates speaking from the famous podium of the UN General Assembly. Findings showed that Palestinian leaders expressed more hope for peace than did leaders from Israel. Circling back to the theories of the politics of hope and skepticism, the chapter ended by offering several explanations for why leaders of the lower-power group adopt the politics of hope while leaders of the high-power group adopt the politics of skepticism.

Chapter 7 began by challenging the intuitive assumption that hope is one of the drivers of political change. Building on research on collective action in the context of climate change, I first provided proof to the contrary. It seems that the extent to which people believe that solving the problems of climate change is possible (the expectation dimension of hope) is unrelated to their support and engagement in environmental activism. Yet the picture in the realm of peace is different. Across studies and samples, both dimensions of hope seem to be highly predictive of a host of conflict-related outcomes. Experimental studies go beyond prediction by establishing a causal link between hope and peace-promoting outcomes. Perhaps most revealingly, when predicting peace-promoting outcomes, the wish dimension of hope seems to matter more than the expectation dimension. More on the hierarchy of the dimensions in the following subsections.

Answering the Research Questions

Three overarching questions guided this book's attempt to decipher how hope functions in the seemingly hopeless conditions of intractable conflict. The first question, "What is hope?" was answered first in order to establish a firm foundation for the rest of the book. Integrating psychological and philosophical research, I suggest that hope can be defined by two components that are both necessary and sufficient for hope to exist: a wish to attain a goal and some expectation that the goal can be attained (Lopez & Snyder, 2003; Sagy & Adwan, 2006; Staats, 1989). Levels of hope rise the more one wishes (i.e., desires, aspires) to attain a goal and the more one expects (estimates, assesses) that the goal will be attained. Chapter 3 provides the theoretical bedrock for conceptualizing and operationalizing hope as a bidimensional construct, while the rest of the book demonstrates the model's utility in a wide variety of contexts.

Conceptualizing hope as a bidimensional construct curbs the inherent problem of hope's confusing nature, sometimes signifying wishes and sometimes expectations. However, the utility of the bidimensional model goes well beyond bypassing a semantic problem. The bidimensional model of hope generated novel, sometimes counterintuitive insights about how hope functions in conflict and enabled new pathways of exploration and comparisons. Chapters 4, 5, 6, and 7 offer numerous examples of the benefits of the bidimensional model. Nonetheless, I feel I have provided only a framework to understand the concept of hope and that further development of the bidimensional model is genuinely needed. Investigating the relationship between the dimensions under different contexts is one example of a new research avenue that could be highly useful. Another path for investigation can examine the relative impact of each dimension on a host of attitudinal and behavioral outcomes. Working with the bidimensional approach merely opened the doors for innovative social science research.

The second overarching research question concerned the determinants of hope during conflict. That is, what, if anything, predicts the level of hope for peace among those involved in conflict? Chapter 4 presented findings from the Hope Map Project that show that age, secularity, and dovish political ideology predict higher hope for peace among those mired in intractable conflict. Broadly speaking, older, secular, and left-wing individuals have more hope for peace than the young, the religious, and those identifying with the political right. Furthermore, people's beliefs in their ability to influence politics predict their expectations (but not wishes) for peace. Stated differently, those with high-efficacy beliefs have higher expectations that peace can materialize than do those with low-efficacy beliefs. This finding suggests that many might understand peace as a bottom-up process rather than an exclusively top-down process that ordinary people cannot control.

Another finding is that the degree to which people are comfortable with uncertainty predicts their wishes (but not expectations) for peace. This implies that the desire for peace necessitates some degree of risk-taking. For those living in conflict for decades, peace is strange and uncertain. Findings suggest that accepting the unpredictable nature of peace creates more space for desiring peace. It spears that people who feel more at ease with uncertainty "allow" themselves to wish for peace more than do those who are less comfortable with uncertainty and unpredictability.

These results shed important light on the demographic, social, and psychological predictors of hope for peace. However, perhaps the most illuminating finding is the difference between the levels of hope for peace among Palestinians and Israelis. Data collected in Israel proper, the West Bank, and the Gaza Strip show that Palestinians and Israelis exhibited similar wishes for peace but that Palestinians' expectations for peace—that is, their belief in the likelihood of a peaceful resolution to the conflict—was much higher than Israelis'. In fact, across all definitions of peace presented to participants, Palestinians' expectations of peace were consistently higher than those of Israelis. Comparing the frequency of hope for peace expressed by the leadership of the two societies at the General Debate of the UN (Chapter 6) revealed a similar picture, albeit one with a significant difference. Averaged across forty-six speeches made at the UN General Assembly, Palestinian leaders expressed more hope for peace than did their Israeli counterparts. This time, Palestinians were higher on the wish dimension, expressing more desires for peace than Israeli representatives.

The first comparison between Israelis' and Palestinians' hope for peace was conducted among the populace; the second was among their leaders. Across the two comparisons, Palestinians' hope for peace was higher than Israelis'. Note that these findings were drawn using two distinct research methodologies, the first from analyzing survey data collected from 1,000 Israelis and Palestinians and the second from linguistic analyses of forty-six speeches made in the UN General Assembly. It is critical to note that, in the survey data, the expectation for peace was the determining dimension driving the difference between Israelis' and Palestinians' hope. In the speeches, the wish dimension was the one producing the difference. Yet, overall, the picture is clear: Palestinians, the lower-power party to the conflict, exhibit more hope for peace than do Israelis.

Two intertwined explanations can potentially account for the higher levels of hope among the disadvantaged group. The first pertains to the different conflict-related realities experienced by members of high- and low-power parties. The reality of the disadvantaged group is saturated with violence and instability, so members of the low-power group are likely to be eager to change the status quo. Moreover, for low-power groups living under military occupation, peace and its assumed fruit—self-determination—is the opposite of the oppressive status quo. The hope for peace is thus bound to be high among the powerless.

In contrast, the status quo might be bearable and even preferred for the high-power group.[1] Though advantaged groups pay some price for the continuation of the conflict, like high expenditures to maintain military superiority, the price is less acute and observable than the one endured by the disadvantaged. Peace, on the other hand, comes with a price for the advantaged in the form of the need to compromise and share power and resources. It thus makes sense that hopes for peace will be lower among advantaged group members compared to disadvantaged group members.

From a historical viewpoint, it seems that Israelis' desires for peace were explicit and unquestionable throughout the first sixty years of the state's existence (Antonovsky & Arian, 1972). However, since the end of the Second Intifada, and especially during the Netanyahu administration, the conflict became much more tolerable for Israelis. Consequently, peace became a less cherished goal. Of course, episodes of violence and hostilities continue to occur and elicit much concern from Israelis when they do. However, the violence directed toward Israelis eventually subsides, giving way to a relatively comfortable life for them. Peace is understandably less urgent when the reality of the conflict is not that devastating, which explains the lower hopes for peace in Israel compared to the OPT.

Leaders, and the political approach they endorse, might be the second reason for the gap between Palestinians' and Israelis' hope. A chief driver of political struggles is hope (Bloch, 1959; Greenaway et al., 2016), so leaders of nations struggling for self-determination are inclined to speak and act according to the guidelines of the politics of hope, creating a vision and committing to achieving it. The politics of skepticism, on the other hand, is

[1] A yet unpublished study I conducted with colleagues in 2022 shows that the status quo is seen as a favorable situation for most Israelis.

not guided by dreams or vision. Instead, it encourages the populace to lower their expectations for the future and concentrate on the here and now. The politics of skepticism is an avenue that may be adopted by leaders who seek to make the status quo permanent.

Chapter 6 describes how and why the Palestinian leadership has been endorsing the politics of hope while the Israeli leadership abandoned the politics of hope at some time during the early 2000s and gradually adopted the politics of skepticism. As presented, one manifestation of the different approaches adopted by each party is the higher number of expressions of the wish for peace in speeches of Palestinian leaders compared to Israeli leaders. The quantitative comparison provides a solid, albeit initial indication that Israeli leaders have turned their back on the politics of hope. Much more needs to be uncovered to better understand and articulate these processes. Furthermore, the political approaches endorsed by the leadership of the two societies are not necessarily stationary. They might be evolving in different ways. One alarming direction is that Palestinians will also lose their hope for peace.

I proposed two explanations for the Israeli-Palestinian "hope gap": one based on the different conflict experiences of low- versus high-power groups and the other on the politics endorsed by their leaders. Publics and leaders operate in the same ecosystem, in this case, the ecosystem of an asymmetrical, intractable conflict, and so these explanations undoubtedly converge and reinforce each other. Leaders of the disadvantaged are inclined to adopt the politics of hope, while leaders of the powerful can afford to adopt the politics of skepticism. At the same time, leaders create political structures that persuade and mobilize their constituents. As such, they can challenge or strengthen it.

The third research question posed at the beginning of the book pertains to the role of hope in conflict resolution processes and, more generally, in driving political change. Does hope matter? Can it help in promoting solutions to large-scale multilayered problems like intractable conflict? Different, sometimes contradicting answers to these questions were provided throughout the book. Here I present the major insights emerging from these answers.

The first, advocated by political thinkers from the Marxist Ernest Bloch (1959) to US President Barak Obama (2006), is that hope is necessary for any movement toward political change. A broader argument posits that hope has been vital in propelling humankind's social advancement and physical survival (Tiger, 1989). Empirical investigation made by psychologists in the past forty years supports this proposition by showing that hope predicts higher achievement (Curry et al., 1997), better general well-being (Cheavens et al., 2005), and even longer life-spans (Lee et al., 2019). Looking more closely at conflict and its transformation, hope for peace was found to be one of the most robust predictors of conciliatory attitudes and behaviors (Leshem & Halperin, 2020b, 2021). Experimental work corroborated these findings by showing the causal link between increased hope for peace and greater support for compromise and peacebuilding (Cohen-Chen et al., 2015; Leshem, 2019; Leshem et al., 2016).

Although strong evidence supports the claim that hope is necessary for human survival, social advancement, or political transformation, the evidence does not prove that hope is a *sufficient* factor for these processes. It is almost an axiom that hope alone cannot explain humankind's progress. The question of the weight of hope vis-à-vis other elemental human conditions (like power) is thus still open. "Hope is not a strategy" is a famous slogan used by business mavericks and military generals. Indeed, hope scholars, including myself, should wonder whether too much burden is mounted on hope's (broad) shoulders. In addition, I have also described the downsides of hope, including hope's tendency to soothe and reduce anxiety (Breznitz, 1986). In this regard, hope's palliative nature can undermine large-scale political changes because people might immerse themselves in hoping for political transformation instead of working to attain it. For the most part, hope is associated with agency, responsibility, and activity (Snyder, 1994). However, facing the arduous task of transforming the rigid structure of power and intractability, even those most committed to change might settle for hoping for the change at the expense of actively engaging in its attainment.

This proposition has emerged in the talks we had with Palestinian and Israeli peacemakers who bravely face these questions on a daily basis (Chapter 5). Not only do they need to endure the setbacks and frustrations innate to peacebuilding work and the insulting attacks aimed at them by their fellow group members, but they also need to face the question of hope. Is their hope for peace realistic? Is it beneficial? Or is hope just a catalyst of burnout and complete despair? From their answers, we have learned an important lesson. Hope is beneficial regardless of whether it produces change. It serves as a compass, a value to be cherished when the whole system seems to go the other way. Peacebuilders hold on to hope despite the unfavorable chances for change. It is a matter of principle and integrity.

The interviews with peacebuilders shed light on another crucial question concerning the relative importance of the two dimensions of hope. What is more important? The extent to which one wishes to attain a goal, or to which one believes attainment is possible? Responses to this question will be partial because the answer depends on circumstances, the character of the hopeful individual or collective, and the nature of the goal.

Notwithstanding these limiting factors, some notions emerged from the book's exploration into hope's dimensions, hinting at a hierarchy between the components. For example, we found no evidence of exceptionally high expectations for peace among peacebuilders. Israeli and Palestinian peacebuilders are hesitant, just like most residents of the region, about the chances for peace. However, their desires for peace were unequivocal and unshaken. From their reflections, we learn that peacebuilders' stubborn engagement in the taxing and mostly unappreciated peacebuilding work derives from their firm desires and aspiration to bring peace to a tormented land, not necessarily their belief that this can be done.

Furthermore, in models where the two dimensions were pitted against each other, the effect of the expectation dimension never exceeded the effect of the wish dimension. The impact of the dimensions on Palestinians' and Israelis' willingness to compromise and act for peace was commonly stronger for the wish dimension. Indeed, it is difficult to draw concrete conclusions from the limited number of studies that compared each dimension's contribution to peace-supporting outcomes. If anything, these partial conclusions demonstrate the importance of examining hope as a bidimensional construct and the need to give scholarly attention to each dimension.

The supremacy of the wish dimension over the expectation dimension resonates with the writings of thinkers and political leaders. Vaclav Havel, the political dissent who became the first democratically elected president of Czechoslovakia, was one of the prominent voices to talk about hope as a notion signifying a wish. "Hope is not the conviction that something will turn out well," explained Havel, "but the certainty that something makes sense, regardless of how it turns out." Aspirations, dreams, visions, and desires drove Havel's hope as he led the Czechoslovakian people to political freedom, not a conviction that these dreams could come true.

Martin Luther King, Jr.'s "I Have a Dream" speech is perhaps the most notable example of the strength of the wish dimension of hope. The speech is remembered for the concrete aspirations and vivid vision it portrays in the form of uncompromising demands for equality and acceptance of the black people of America. King promises that these demands will ultimately be met, so the expectation dimension of hope is also evoked. However, the promises are overshadowed by the power of the "Dream," a dream King and his listeners, then and today, desire and aspire for. Building on philosophical deliberations and empirical findings presented in this book, one of the insights I cautiously propose is that, when it comes to hope for peace, the wish dimension has primacy over the expectation component. I am still unsure if this primacy is an issue of order or impact and whether this observed primacy holds for hope for political transformation in general. Indeed, this proposition deserves all the scrutiny and criticism it can get. I am the first to admit that this claim might depend on many factors, including the nature of the hoping actors and hoped-for political change. The hierarchy of the dimensions is a topic worthy of further debate and exploration.

Theoretical and Methodological Contributions

The first contribution I sought to make to the theoretical explorations of hope amidst conflict is dismantling the damaging boundaries separating philosophical and psychological inquiries into hope. The two commonly isolated approaches have provided valuable insights into the structure and meaning of hope and how it may function during conflict. Combining the disciplinary approaches was aimed at making our understanding of hope more comprehensive and holistic. Nevertheless, integrating the disciplines provided more than just a broader account of hope. It helped raise questions that were not asked and provided answers to theories never tested.

For instance, by amalgamating the disciplines, the book was able to shed light on the negative ramifications that hope might have during conflict. Conducted primarily by psychologists, previous studies on hope during conflict assumed that hoping for peace is a "good thing" that should be nurtured and cultivated in the challenging conditions of conflict (Leshem & Halperin, 2020a; Rosler et al., 2017; Sagy & Adwan, 2006). Indeed, the positive qualities of hope are intuitive and known. However, philosophical debates outlined in this book also exposed hope's dangers and pitfalls and thus raised the necessity of more boldly reexamining hope's role in conflict. Exposing hope's palliative aspects that promote psychological well-being but hamper political transformation is one of the fruits of revisiting the axiom that hope is always advantageous.

Connecting the two disciplines also allowed me to depart from narrow disciplinary jargon and move to arguments and terminologies that a larger array of readership can enjoy. This move, I believe, may benefit students, teachers, and researchers with backgrounds in conflict resolution, peace and justice studies, or other interdisciplinary scholars from the social sciences and humanities. Furthermore, a complex phenomenon like intractable conflicts requires multiple lenses if we seek to understand and transform these situations. Integrating philosophical and psychological approaches is the first step in aggregating insights from different schools of thought on the same subject matter. This book provides a glimpse into the advantages of incorporating insights from two disciplines and, most importantly, offers novel insights on hope amidst conflict that could only emerge from a combination of approaches and perspectives.

This book's second contribution is the introduction and systematic use of the bidimensional model of hope. Conceptualizing hope as a combination of two core components—wishes and expectations—has been utilized before by me and others (e.g., Leshem, 2017, Lopez & Snyder, 2003; Sagy & Adwan, 2006). However, providing a solid theoretical framework for the bidimensional conceptualization and arranging it into a unified model that can be efficiently utilized in various contexts is novel and beneficial for all those interested in hope. I devote an entire chapter to the theoretical rationale behind the bidimensional model and to demonstrating its empirical utility. In the remainder of the book, I applied the model using diverse methods like quantitative inferential analyses, linguistic discourse analyses, and qualitative interviews.

One of the essential features of the bidimensional model is its intuitiveness and simplicity. The model is intuitive because a quick, unsophisticated inspection of the meaning of hope reveals its core components. It is simple because its implementation is user-friendly in any context in which one seeks to study hope. It is my conviction that the confusing way the word "hope" is being used in everyday life (sometimes to signify wishes and sometimes expectations) made it difficult to start a robust scholarly exploration of hope during conflict and led astray much of the empirical research on hope conducted in conflict zones. Recognizing that hope has two distinct meanings corresponding to two distinct dimensions paves the way to using the bidimensional model in almost any context.

Introducing the Hope Map Project with its aims, methods, and initial results is the third contribution this book makes. The Hope Map Project measures the levels of hope for peace among people mired in conflict and identifies its demographic, psychological, and sociopolitical correlates. As presented in this book and elsewhere (Leshem & Halperin, 2020b, 2021), the first phase of the Hope Map Project conducted in Israel-Palestine in 2017 produced an impressive volume of findings about the hope for peace in intractable conflict. However, the project intended to go beyond a one-time, one-region study and expand in scope, frequency, and geographic spread.

With this in mind, three projects that extend the original 2017 study were recently conducted, one among Turkish and Greek Cypriots (N = 1,100) and another two in Israel, the West Bank, and Gaza (N = 2,600). Some of the aims of the Cyprus Hope Map were to test whether findings from Israel-Palestine replicate in a low-intensity conflict and investigate new correlates of hope, such as the strength of ethnic identification. The goals of the 2022 and 2023 Israel-Palestine Hope Map were to identify changes in hope across time and expand the targets of hope to domains other than peace, including the hope to defeat the adversary. Analyses of the data were not completed in time to be included in the book. Nevertheless, initial findings seem to add considerable depth and breadth to our understanding of hope's role during conflict. In the long run, the Hope Map Project is intended to become an ongoing longitudinal study administered at regular intervals in conflict zones worldwide to provide the most comprehensive account of hope amidst conflict.

Applied Implications

I am convinced that placing hope at the center of attention of conflict resolution and peacebuilding practice can advance conflict transformation in various ways. Perhaps the most immediate is to actively promote and circulate the idea of hope for peace into the discourse of societies involved in conflicts. As described in Chapter 7, inducing hope for peace increases people's support for concession-making and peacebuilding programs (Cohen-Chen et al., 2015; Leshem, 2019). Although the interventions were tested only in controlled experimental environments, their essence can be adapted into real-world media campaigns disseminated in conflict zones via traditional and new media channels. An Israeli NGO is currently developing one such campaign to raise Jewish Israelis' support for peace. Building on the finding that members of the adversarial group, although untrusted and disliked, are effective hope-inducing messengers (Leshem, 2019), the campaign's protagonists are Palestinian bloggers.

Inserting the idea of hope for peace into the conversation in Israel-Palestine comes at a time when the notion of hope is absent from public and international discourse regarding the conflict. Even left-wing politicians have been abstaining from voicing hope for peace. Because domestic and geopolitical conditions are unfavorable for conflict transformation in the region, hope for peace has become tantamount to naïvete and detachment. Once a noble idea, hope is treated nowadays with sarcasm and ridicule. Still, the book's exploration into hope has also highlighted that hope is not just about the possibility of peace but also about the desire for it. So even when expectations are low, dovish leaders must clearly and deliberately voice their desires and aspirations for peace, perhaps in the same way Martin Luther King, Jr., declared his desires for equality and freedom as a tool for political change.

The deliberate use of hope can also benefit the peacebuilding community. Working in the discouraging conditions of conflict, peacebuilders and activists commonly experience burnout and fatigue (Vandermeulen et al., 2022). Learning to nurture and sustain hope can decrease weariness and invigorate activism. Recently, I have started to lecture and conduct workshops on hope outside academia, mostly for groups working for social and political change in their communities and elsewhere. In these events, we discuss hope's benefits and drawbacks, the bidimensional model, and hope's determinants and potential outcomes. From their feedback I have discovered that learning and deliberating on hope has been meaningful for participants as individuals and involved citizens and that these workshops restored their confidence in and commitment to their goals. Furthermore, in these talks, I propose that the type of hope most sustainable and effective in driving social and political transformation needs to be modest in expectation but bold in the level of wishes. I expand on this proposition next.

Conclusion: Exploring the idea of Optimal Hope

One of the central arguments advocated throughout this book is that dividing hope into its two dimensions is critical to understanding the concept. The bidimensional model helps define hope, measure it, and reveal its correlates. It also draws attention to how people talk about hope, whether these people are national leaders speaking at the UN or local grassroots peacebuilders talking about their activism. The theoretical and methodological utility of the bidimensional approach was demonstrated in all previous chapters. I believe that dividing hope into wishes and expectations might also benefit people in their continuous attempt to pursue goals and avoid adversity.

Actively distinguishing between the two dimensions is advantageous because their fusion creates two undesirable conditions. First, high wishes for a goal will likely generate unduly high expectations of fulfillment. This "wishful thinking" process might lead the wishful person to portray a rosy picture of reality and downplay the challenges that lie ahead. In the long run, this type of hope is not sustainable because harsh reality and challenging circumstances will ultimately burst the bubble of naïvete. The second problem with fusing the dimensions comes from the opposite end, when low expectations for attaining a goal deflate the wishes for attainment. In these circumstances, people's low beliefs in the possibility of achieving the goal suppress their wishes for the goal. Both processes are natural: people tend to exaggerate the likelihood of attaining things they want badly and stop wishing for things that are hard to attain. However, both processes create stagnation and are, therefore, detrimental to political transformation.

The undesired fusion of wishes and expectations is likely to occur in extreme conditions like protracted conflict, where hope is constantly challenged. Some peacebuilders might be inclined to generate a favorable assessment of the likelihood of peace to match their strong desires for it. Most people, however, will tend to suppress their wishes for peace to align with their low beliefs in the possibility of resolution. This process is typical in wide sectors of Israeli and Palestinian society, where people gave up on wishing for peace because the chances of success are so low.

The question is how to avoid this fusion and what can be suggested instead. I propose an optimal type of hope that tackles the inclination to merge wishes and expectations in working for social and political transformation. Optimal hope is based on actively and deliberately disaggregating the two dimensions of hope in any thoughts concerning political transformations, including the transformation of conflicts. The consequence of this deliberate disaggregation creates two qualities fundamental to optimal hope. The first is an uncompromising analysis of reality. This analysis seeks to detect, rather than ignore, the toughest challenges that exist and lie ahead in the struggle for peace, justice, or equality, including the unpleasant costs of its pursuit.

The second quality deriving from the active distinction between expectations and wishes is that extreme obstacles and unfavorable odds do not diminish the wish for the goals. The wishes stay high, unequivocal, and unshaken, not even by adversity and setbacks. In line with Vaclav Havel's interpretations of hope (Havel, 1990), optimal hope is driven chiefly by hope's wish dimension rather than the expectation dimension. Interestingly, this type of hope emerged in the thoughts and ideas expressed by some activists interviewed for this book. These activists insisted that their hopes were not based on an assessment that peace is achievable. They admit they would have no hope if it were the case. Their evident hope for peace, it seems, was spurred by an uncompromising wish or peace and a feeling of responsibility to generations to come.

Some applications of optimal hope can be made to the fields of peacebuilding and conflict resolution. First, optimal hope refutes the claim that discussing peace during intractable ethnonational conflicts is unwarranted. Indeed, the reality of the conflict, with its constant presence of hostilities, fatalities, and existential threats, is likely to engender a skeptical outlook on the future of peace, not an optimistic one (Maoz & Shikaki, 2014). Therefore, discussing hope for peace during intractable conflict is often considered a waste of time or even a dangerous delusion. Those using optimal hope are neither naïve nor delusional. They are well aware of the depths of the conflict and the incredible obstacles in the way of peace. There is no need to minimize the challenges or portray a brighter picture of reality because these will not deter the optimal hopers' desires and passions for peace that serve as the chief drivers of their hope.

Furthermore, at least partially, optimal hope protects from despair, frustration, and anguish when hopes for peace are dashed. The main concern with promoting hope for peace is that promises that cannot be kept should not be made. However, optimal hope does not require promises. As a construct founded more on aspirations and desires for a goal than on the possibility of achieving it, optimal hope is more adaptable to adversity, more flexible in the face of changing circumstances, and thus more sustainable throughout long struggles. In other words, because optimal hope is less dependent on expectations, it is more enduring in the face of hardship and more immune to despair. The driving forces of optimal hope—unshaken desires, aspirations, and wishes for social and political change—are based on internal motivation rather than external assessments. Optimal hope is thus less likely to be abandoned. Visit my website for more thoughts on optimal hope. https://www.rethink-hope.com/

***

I end with two quotes highlighting the need for hope in intractable conflicts. The first is by Stav Shafir, who was elected to the Israeli Parliament in 2013 at the age of twenty-seven. The second is from Raef Zreik, a Palestinian lawyer and academic known for his work on the conflict. Both quotes call our attention to the role of hope as a catalyst for change.

> We were born into a very cynical state, a somewhat damaged society in a state of despair, and what we have to do is to bring a genuine feeling of hope and a commitment to realize it. That's our responsibility. It's a responsibility that we'll take from the street and transfer to the Knesset. (Schenker, 2013)

> My optimism does not derive from a feeling that we can decipher history's secret plan or hasten its unfolding. My optimism is humbler. It is not deduced from a clear analytical conclusion but from historical human experiences. Our experience as humans has taught us that sometimes, only sometimes, there are historical narratives that end well. Our experience has also taught us that these positive endings were not coincidental. Rather, there were those who diligently worked to make them happen, though luck had an important role as well. What is certain is

that we need to move forward and move fast so that luck will be on our side. And remember: though we can never be certain in our success, there is also no certainty in our failure. (Zreik, personal note)

References

Antonovsky, A., & Arian, A. (1972). *Hope and fears of Israelis*. Jerusalem Academic Press.

Bloch, E. (1959). *The principle of hope* (vol. 1–3). MIT Press.

Breznitz, S. (1986). The effect of hope on coping with stress. In M. H. Appley (Ed.), *Dynamics of stress* (pp. 295–306). Plenum.

Cheavens, J., Michael, S. T., & Snyder, C. R. (2005). *The correlates of hope: Psychological and physiological benefits* (J. Eliott, Ed.; pp. 119–132). Nova Science.

Cohen-Chen, S., Crisp, R. J., & Halperin, E. (2015). Perceptions of a changing world induce hope and promote peace in intractable conflicts. *Personality and Social Psychology Bulletin*, *41*(4), 498–512. https://doi.org/10.1177/0146167215573210

Curry, L. A., Snyder, C. R., Cook, D. L., Ruby, B. C., & Rehm, M. (1997). Role of hope in academic and sports achievement. *Journal of Personality and Social Psychology*, *73*, 1257–1267.

Descartes, R. (1649/2015). *The passions of the soul and other late philosophical writings*. Oxford University Press.

Gravlee, G. S. (2020). Hope in ancient Greek philosophy. In S. C. van den Heuvel (Ed.), *Historical and multidisciplinary perspectives on hope* (pp. 3–23). Springer International. https://doi.org/10.1007/978-3-030-46489-9_1

Greenaway, K. H., Cichocka, A., van Veelen, R., Likki, T., & Branscombe, N. R. (2016). Feeling hopeful inspires support for social change: Hope and social change. *Political Psychology*, *37*(1), 89–107. https://doi.org/10.1111/pops.12225

Hatz, S. (2019). Israeli demolition orders and Palestinian preferences for dissent. *Journal of Politics*, *81*(3), 1069–1074. https://doi.org/10.1086/703211

Havel, V. (1990). *Disturbing the peace: A conversation with Karel Hvížďala*. Vintage.

Heraclides, A. (2011). The Cyprus Gordian knot: An intractable ethnic conflict. *Nationalism and Ethnic Politics*, *17*(2), 117–139. https://doi.org/10.1080/13537113.2011.575309

Lee, L. O., James, P., Zevon, E. S., Kim, E. S., Trudel-Fitzgerald, C., Spiro, A., Grodstein, F., & Kubzansky, L. D. (2019). Optimism is associated with exceptional longevity in 2 epidemiologic cohorts of men and women. *Proceedings of the National Academy of Sciences*, 201900712. https://doi.org/10.1073/pnas.1900712116

Leshem, O. A. (2017). What you wish for is not what you expect: Measuring hope for peace during intractable conflicts. *International Journal of Intercultural Relations*, *60*, 60–66. https://doi.org/10.1016/j.ijintrel.2017.06.005

Leshem, O. A. (2019). The pivotal role of the enemy in inducing hope for peace. *Political Studies*, *67*(3), 693–711. https://doi.org/10.1177/0032321718797920

Leshem, O. A., & Halperin, E. (2020a). Hope during conflict. In S. C. van den Heuvel (Ed.), *Historical and multidisciplinary perspectives on hope* (pp. 179–196). Springer International. https://doi.org/10.1007/978-3-030-46489-9_10

Leshem, O. A., & Halperin, E. (2020b). Hoping for peace during protracted conflict: Citizens' hope is based on inaccurate appraisals of their adversary's hope for peace. *Journal of Conflict Resolution*, *64*(7–8), 1390–1417. https://doi.org/10.1177/0022002719896406

Leshem, O. A., & Halperin, E. (2021). Threatened by the worst but hoping for the best: Unraveling the relationship between threat, hope, and public opinion during conflict. *Political Behavior*, 20. https://doi.org/10.1007/s11109-021-09729-3

Leshem, O. A., Klar, Y., & Flores, T. E. (2016). Instilling hope for peace during intractable conflicts. *Social Psychological and Personality Science*, *7*(4), 303–311. https://doi.org/10.1177/1948550615626776

Lopez, S. J., & Snyder, C. R. (2003). Hope: Many definitions, many measures. In S. Lopez & C. R. Snyder (Eds.), *Positive psychological assessment* (pp. 91–108). American Psychological Association.

Maoz, I., & Shikaki, K. (2014). *Joint Israeli Palestinian poll, December 2014*. The Harry S. Truman Research Institute For the Advancement of Peace, The Hebrew University of Jerusalem.

Nietzsche, F. (1878/1974). *Human all too human*. Gordon Press.

Obama, B. (2006). *The audacity of hope: Thoughts on reclaiming the American dream*. Random House Large Print.

Rosler, N., Cohen-Chen, S., & Halperin, E. (2017). The distinctive effects of empathy and hope in intractable conflicts. *Journal of Conflict Resolution*, *61*(1), 114–139. https://doi.org/10.1177/0022002715569772

Sagy, S., & Adwan, S. (2006). Hope in times of threat: The case of Israeli and Palestinian youth. *American Journal of Orthopsychiatry*, *76*(1), 128–133. https://doi.org/10.1037/0002-9432.76.1.128

Schenker, H. (2013). It's our generation's responsibility to bring a genuine feeling of hope. *Palestine-Israel Journal of Politics, Economics & Culture*, *18*(4), 101–107.

Snyder, C. R. (1994). *The psychology of hope*. Free Press.

Spinoza, B. (1677/2000). *Ethics* (G. H. Parkingson, Trans.). Oxford University Press.

Staats, S. R. (1989). Hope: A comparison of two self-report measures for adults. *Journal of Personality Assessment*, *53*(2), 366–375.

Tiger, L. (1989). *Optimism, the biology of hope*. Simon & Schuster.

Vandermeulen, D., Hasan Aslih, S., Shuman, E., & Halperin, E. (2022). Protected by the emotions of the group: Perceived emotional fit and disadvantaged group members' activist burnout. *Personality and Social Psychology Bulletin*, 01461672221092853. https://doi.org/10.1177/0146167222109285

APPENDIX

Publications on Hope During Conflict

Below is a list of twenty-six publications that used measurement tools to gauge hope during conflict. Only quantitative studies are included in the list. As the list reveals, the conceptualization and operationalization of hope differ significantly from study to study. More alarmingly, in many studies, it is difficult to determine how participants interpreted the word "hope." Did they think they were asked about their "wishes," "expectations," or some mixture of both? I tried to provide an educated guess about the probable interpretation of hope. In some studies, this task was relatively intuitive, while estimating the likely interpretation was impossible in others. In addition, it seems that, in some studies, the researchers focused and reported on a specific dimension (still referring to it as "hope") but that participants interpreted hope based on the other dimension. Another problem is that, in some studies, how hope was operationalized does not derive from how it was conceptualized.

(W) = hope as wish, (E) = hope as expectations, (?) = unclear

Authors	Years	Title	Conceptualization	Operationalization	Nationality of sample	Comments
Antonovsky & Arian	1972	Hope and Fears of Israelis	No definition of hope	Self-Anchoring Striving scale based on Hadley Cantril; "What are your wishes and hopes for the future of our country? If you picture the future of Israel in the best possible light, how would things look, let us say, ten years from now?" (W) "Looking at the ladder, suppose your greatest hopes for Israel are at the top; your worst fears at the bottom. . . . Where do you think Israel will be on the ladder five years from now?" (E)	Jewish Israelis	People are asked to describe their ideal of Israel's future first, which might elicit more positive, wishful answers

Authors	Years	Title	Conceptualization	Operationalization	Nationality of sample	Comments
Staats & Partlo	1993	A Brief Report on Hope in Peace and War, and in Good Times and Bad.	Hope is defined as an interaction between wishes and expectations (W; E)	The items are not provided, but from the results it is perceivable that the operationalization distinguishes between wish and expectation (W; E)	US students and parents	The distinction between wish and expectation dimensions is explicitly made in both conceptualization and operationalization
Moeschenberger et al.	2005	Forgiveness in Northern Ireland: A Model for Peace in the Midst of the "Troubles"	Hope is referred to as a "trait characteristic" but is not further conceptualized (?)	No appendix was found, but it seems that questions measuring "goal-directed thinking" are used to derive individuals' hopefulness (?)	Catholic, Protestant, and other residents of Northern Ireland	Lack of conceptualization and operationalization
Sagy & Adwan	2006	Hope in Times of Threat: The Case of Israeli and Palestinian Youth	Hope is defined as "the interaction between wishes and positive future expectations" (W; E)	Participants were asked to rate independently how much they wished for a specific future occurrence and the extent of their expectation for it to occur (W; E) Expectation value × wish value = hope value	Jewish Israelis, Palestinians of the OPT	The distinction of wish and expectation is fulfilled both in the conceptualization and the operationalization
Halperin et al.	2008	Emotions in conflict: Correlates of fear and hope in the Israeli-Jewish society	Hope is defined as "an integrated reaction that consists of cognitive elements, including expecting and planning a positive occurrence with positive affect." (E)	Personal and collective hope are distinguished: "To what extent do you hope that . . ." (?) "Visiting Damascus as a tourist." "Comprehensive peace" "Cooperation with Arab countries"	Jewish Israelis	Some participants might have reported their wishes for these propositions, while others reported their expectations.

Halperin & Gross	2011	Emotion regulation in violent conflict: Reappraisal, hope, and support for humanitarian aid to the opponent in wartime	Hope is defined as an emotion that "involves expectation and aspiration for a positive goal, as well as positive feelings about the anticipated outcome" (E; W)	Participants were asked to rate the extent to which the recent events made them feel hope regarding the future of the conflict (?)	Jewish Israelis	The conceptualization of hope includes the two dimensions, but the operationalization does not. It is thus hard to know what dimension was gauged.
Nasie & Bar-Tal	2012	Sociopsychological Infrastructure of an Intractable Conflict Through the Eyes of Palestinian Children and Adolescents	Hope is defined as "awaiting or wishing for realization of a concrete objective, including the aspiration to be extricated from negative conditions" (E or W)	Hope is coded according to the conceptualization, and "expressions that reflect hope in the writings included descriptions of longing for freedom, victory, liberation from the occupation, return of refugees to their homeland, emancipation of Jerusalem, avenging the blood of the *shahids*, and yearning for a bright future." (?)	Palestinians from the OPT (including Gaza)	Hope is conceptualized as either expectation or wish. However, since the coding framework is not given, the operationalization cannot be fully reconstructed. It does appear that the paper focuses on the wish dimension due to the use of words such as "longing" and "yearning."
Cohen-Chen et al.	2014	Hope in the Middle East: Malleability Beliefs, Hope, and the Willingness to Compromise for Peace	Hope is defined as a positive emotion that is activated "when one visualizes a meaningful goal of which there is intermediate probability of achievement, followed by a positive change in mental state" (E)	"I am hopeful regarding the end of the Israeli-Palestinian conflict" (?), "I don't expect ever to achieve peace with the Palestinians" (E), and "There is no use in really trying to end the conflict because it probably won't happen" (E) "With regard to the Israeli Palestinians conflict, what has been will always be, and the conflict will stay this way forever" (E)	Jewish Israelis	The conceptualization lacks the wish dimension (though it is implicit in the term "meaningful" and, aside from the first item, the operationalization appears to focus solely on the expectation dimension

Authors	Years	Title	Conceptualization	Operationalization	Nationality of sample	Comments
Cohen-Chen et al.	2014	The Differential Effects of Hope and Fear on Information Processing in Intractable Conflict	Hope is defined as a "secondary, highly cognitively-based emotion which involves expectation and aspiration for a positive goal in the future" (as in Staats & Stassen, 1985; Stotland, 1969) (E, W)	To which extent do you feel the emotion of "hope regarding the future of Israeli-Palestinian relations" (?)	Jewish Israelis	The conceptualization includes both dimensions, but the measurement does not. Thus, some participants might have based their answers on the wish dimension and some on the expectation dimension.
Cohen-Chen, Crisp & Halperin	2015	Perceptions of a Changing World Induce Hope and Promote Peace in Intractable Conflicts	Hope is defined as "a discrete emotion manifested by a forward-oriented cognitive appraisal of a situation as improving in the future" (E)	"When I think about the future of the conflict, I imagine a situation which is better than now." (E) "I am hopeful regarding the end of the Israeli-Palestinian conflict." (?) "Israel might as well give up because it cannot resolve the conflict." (E) "I don't expect ever to achieve peace with the Palestinians." (E) "Under certain circumstances and if all core issues are addressed, the Israeli-Palestinian conflict's nature can be changed." (E)	Jewish Israelis	In line with the conceptualization, items one, three, four, and five aim at expectations. However, item two is unclear, and so participants might have reported either their wishes or expectations for a "better future."

Cohen-Chen et al.	2016	Hope Comes in Many Forms Out-Group Expressions of Hope Override Low Support and Promote Reconciliation in Conflicts	Hope is defined as "a positive emotion that arises due to a cognitive process involving imagining a desired future." (W)	"To what extent do you feel hopeful regarding peace in the future?" (?) "In light of the Palestinian response, to what extent did this outline lead you to experience hope?" (?)	Jewish Israelis	The phrasing in the conceptualization appears closer to the wish dimension, as the term "desired" is used, whereas the operationalization seems to aim at the expectation dimension more than the wish dimension. Still, participants might have interpreted the question in several ways.
Leshem, Klar & Flores	2016	Instilling Hope for Peace During Intractable Conflicts	Hope is defined as "comprised of a wish to attain some goal . . . and the positive but not certain expectations to attain it" (W, E)	Two dependent variables related to hope were measured: Hopefulness: "To what extent do you agree with the statement 'When I saw the clip, I felt hopeful' " (?) Belief in the likelihood of resolution: "To what extent do you believe that 'both sides of the conflict can share the same land; neither side will be able to compromise in the coming future' " (E)	Jewish Israelis	The conceptualization distinguishes between the two elements; however, the operationalization does not do so explicitly and focuses solely on the expectation dimension.

Authors	Years	Title	Conceptualization	Operationalization	Nationality of sample	Comments
Haroz et al.	2017	Measuring Hope Among Children Affected by Armed Conflict: Cross-Cultural Construct Validity of the Children's Hope Scale	Hope is defined as "the process of thinking about one's goals, along with the motivation to move toward and the way to achieve those goals" (as in Snyder, 1995, p. 355) (?)	Children's Hope Scale (Snyder, 1997) "I think I am doing pretty well," "I am doing just as well as other kids my age," "I think the things I have done in the past will help me in the future," "I can think of many ways to get the things in life that are most important to me," "When I have a problem, I can come up with lots of ways to solve it," "Even when others want to quit, I know that I can find ways to solve the problem."	Children from Burundi, Nepal, Indonesia	The dimensions of hope are not distinguishable, neither in the conceptualization nor in the operationalization.
Leshem	2017	What you wish for is not what you expect: Measuring hope for peace during intractable conflicts	"Hope requires both wishes and expectations to attain a goal which together elicit the affective dimension of hope" (W, E)	To what extent to do you "Wish the following definitions [of peace] to materialize?" (W) "Expect the following definitions [of peace] to materialize?" (E)	Jewish Israelis	Expectation and wish are distinguished both in conceptualization and operationalization.
Rosler, Cohen-Chen & Halperin	2017	The Distinctive Effects of Empathy and Hope in Intractable Conflicts	Hope is defined as "a highly cognitive-based emotion that involves expectation and aspiration for a positive goal in the future, as well as positive feelings about the anticipated outcome" (E)	To what extent do participants feel "Hope regarding the future of Israeli-Palestinian relations" (?)	Jewish Israelis	The conceptualization does not recognize the wish dimension, and the operationalization appears to focus on the expectation dimension.

Shani & Boehnke	2017	The effect of Jewish-Palestinian mixed-model encounters on readiness for contract and policy support	Hope is defined as "positive feelings resulting from the expectation that positive goals are about to materialize, and from the belief that successful pathways to achieving these goals are available" (E)	Only one out of four items are given "In the future, there will be friendly relations between Arabs and Jews in Israel" (E)	Jewish and Palestinian residents of Israel	Hope is conceptualized and operationalized based solely on the expectation dimension. However, the operationalization cannot be reconstructed as no appendix is available.
Cohen-Chen & Van Zomeren	2018	Yes we can? Group efficacy beliefs predict collective action, but only when hope is high	Hope is defined as the emotional experience "elicited by the cognitive appraisal that a meaningful goal is possible to achieve in the future" (E)	"I feel hope regarding the possibility of resolution of the Israeli-Palestinian conflict" (?) "Under certain circumstances, and if all core issues are addressed, the conflict can be resolved in the future" (E) "It is clear to me that attempts to resolve the conflict are without hope" (?) "Israel should stop trying to resolve the conflict because it is impossible" (E – **double-barreled**) "I don't expect ever to achieve peace with the Palestinians" (E)	Jewish Israelis	Both in the conceptualization and the operationalization, it seems that the authors focused on the expectation dimension of hope.
Goldenberg et al.	2018	Testing the impact and durability of a group malleability intervention in the context of the Israeli-Palestinian conflict	No definition given	"In general, when you think about the Palestinians, to what extent do you feel the following emotions?" "1. Hope regarding the relationship with the Palestinians. (?) 2. Desperation regarding the relationship with the Palestinians. (E) 3. Optimism about the future relationship with the Palestinians." (E)	Jewish Israelis	Hope is not conceptualized, and the operationalization does not distinguish between the expectation and wish dimensions. It seems that most items are focused on expectations, but it is hard to be sure what participants thought they were asked.

Authors	Years	Title	Conceptualization	Operationalization	Nationality of sample	Comments
Cohen-Chen et al.	2019	Dealing in hope: Does observing hope expressions increase conciliatory attitudes in intergroup conflict?	Hope is conceptualized as consisting of three dimensions: "a wish or desire for conflict resolution, a belief that this future is possible, and positive affect prompted by the prospect of resolving the conflict" (W; E)	To what extent do you agree with the statements: "I am hopeful that this conflict will be peacefully resolved in the future" (?), "When I think about the future of the relations between us and the government, I feel hope" (?) "Under certain circumstances and if all core issues are addressed the students' situation can improve in the future" (E) "We should stop trying to resolve this conflict because it will never happen" (E) "I don't expect ever to resolve this conflict" (E) and "To what extent do you feel hopeful in light of the leadership's reaction" (?)	UK students	The conceptualization distinguishes between the dimensions of hope and expectations, yet this is not found in the operationalization, which focuses mainly on the expectation dimension. It is hard to know how participants interpreted hope in some questions.
Elad-Strenger et al.	2019	Facilitating Hope Among the Hopeless: The Role of Ideology and Moral Content in Shaping Reactions to Internal Criticism in the Context of Intractable Conflict	No definition given	"I am hopeful regarding the prospects of improving the relationship between Jewish Israelis and Palestinian citizens of Israel" (?) "I am optimistic regarding the future of the relations between Jewish Israelis and Palestinian citizens of Israel" (E)	Jewish Israelis	Hope is not conceptualized, and the items measured are rather vague. The second item measures optimism, which might be closest to the expectation dimension.

Hasan-Aslih et al.	2019	A Darker Side of Hope: Harmony-Focused Hope Decreases Collective Action Intentions Among the Disadvantaged	Hope is defined as "the emotional experience associated with the desire for improving existing conditions," reflecting "a belief in the possibility of positive change" (W; E)	"Hope for a better future in relation between Arabs/Palestinians and Jews. . . ." (W) "Hope for promoting the status of Arab/Palestinian citizens in the country" (W)	Palestinian citizens of Israel	While the conceptualization contains both dimensions of hope, the operationalization does not. Examining the article, hope was probably interpreted by participants as wish.
Leshem	2019	The Pivotal Role of the enemy in Inducing Hope for Peace	Hope is defined as containing three elements, the wish to attain a goal, the expectation that the goal can be achieved, and active commitment to attain the goal (W; E)	How much do you wish for/ expect the propositions to materialize: (W; E) (1) "Achieving peace as you define and understand it," (2) "Achieving a mutually agreed upon accord that ensures the interests of both peoples," (3) "Achieving a mutually agreed upon accord ensuring independence and freedom for Palestinians and security and safety for Israelis."	Jewish Israelis	Expectation and wish are distinguished both in conceptualization and operationalization.
Hasan-Aslih et al.	2020	The Quest for Hope: Disadvantaged Group Members Can Fulfil their Desire to Feel Hope, but Only When They Believe in Their Power	Hope is defined as "an emotion that arises from a strong desire to be in a different situation than at present," which entails "the appraisal that a desired change is possible in the future" (W; E)	"To what extent do you feel the following emotions," which included "hope for ending the occupation" (?) and "hope for coexistence" (?)	Palestinians of the OPT	The conceptualization distinguishes between wish and expectation. It is difficult to know if participants reported their wishes to "end the occupation" / coexistence" or their expectation that these events will happen (or some amalgam of both).

Authors	Years	Title	Conceptualization	Operationalization	Nationality of sample	Comments
Leshem & Halperin	2020a	Hope During Conflict	Hope is defined as "a combination of two factors, a wish (i.e., desire) to attain peace and some expectation (i.e., assessment of likelihood) that peace can be attained" (W; E)	See Leshem 2017, 2019	Jewish Israelis; Palestinians from the Westbank, and Gaza	Expectation and wish are distinguished both in conceptualization and operationalization.
Leshem & Halperin	2020b	Hoping for Peace during Protracted Conflict: Citizens' Hope Is Based on Inaccurate Appraisals of Their Adversary's Hope for Peace	Hope is defined as "an amalgam of two factors, a desire (i.e., wish) to attain a goal and some expectation (i.e., assessment of likelihood) that the goal can be attained" (W; E)	See Leshem 2017, 2019	Jewish Israelis; Palestinians from the Westbank, and Gaza	Expectation and wish are distinguished both in conceptualization and operationalization.
Leshem & Halperin	2021	Threatened by the Worst but Hoping for the Best: Unraveling the Relationship Between Threat, Hope, and Public Opinion During Conflict	Hope is defined as comprising of two dimensions: "the wish to achieve a goal and the expectation (though not certainty) that the goal can be achieved" (W; E)	See Leshem 2017, 2019	Jewish Israelis; Palestinians from the Westbank	Expectation and wish are distinguished both in conceptualization and operationalization.

Index

For the benefit of digital users, indexed terms that span two pages (e.g., 52–53) may, on occasion, appear on only one of those pages.

Tables and figures are indicated by *t* and *f* following the page number